FIAT 500
Owner's Workshop Manual

by J.H.HAYNES

Associate member of the Guild of Motoring Writers

and J.C.LARMINIE

Models covered

479 cc	FIAT New 500	1957 - 1960
	FIAT 500 Sport	1958 - 1960
499.5 cc	FIAT 500D	1960 - 1965
	FIAT 500F	1965 - 1972
	FIAT 500L	1968 - 1972
	FIAT 500 Station Wagon (Autobianchi Giardiniera Panoramica)	1960 - 1972
594 cc	FIAT 500	1972 onwards

SBN 900550 90 2

 J H Haynes and Company Limited 1973

ABCDE
FGHIJ
KLMNO
PQ*

HAYNES PUBLISHING GROUP
SPARKFORD YEOVIL SOMERSET ENGLAND
distributed in the USA by
HAYNES PUBLICATIONS INC
861 LAWRENCE DRIVE
NEWBURY PARK
CALIFORNIA 91320
USA

Acknowledgements

The greatest help possible has been given by FIAT. Many of the drawings in this book are theirs.

Their London publicity agents, Woolf, Laing, Christie and Partners, and FIAT (England) at Brentford have been most co-operative as have many local FIAT garages.

Brian Horsfall stripped down our own car, with Les Brazier taking the photographs, and Tim Parker edited the text and illustrations.

About this manual

The aim of this book is to help you get the best value from your car. It can do so in two ways. First it can help you decide what work must be done, even should you choose to get it done by a garage; the routine maintenance and the diagnosis and course of action when random faults occur. But it is hoped that you will also use the second and fuller purpose by tackling the work yourself. This can give you the satisfaction of doing the job yourself. On the simpler jobs it may even be quicker than booking the car into a garage and going there twice, to leave and collect it. Perhaps most important, much money can be saved by avoiding the costs a garage must charge to cover their labour and overheads.

The book has drawings and descriptions to show the function of the various components so that their layout can be understood. Then the tasks are described and photographed in a step by step sequence so that even a novice can cope with complicated work. Such a person is the very one to buy a car needing repair yet be unable to afford garage costs.

The jobs are described assuming only normal spanners are available, and not special tools. But a reasonable outfit of tools will be a worthwhile investment. Many special workshop tools produced by the makers merely speed the work, and in these cases guidance is given as to how to do the job without them, the oft quoted example being the use of a large hose clip to compress the piston rings for insertion in the cylinder. But on a very few occasions the special tool is essential to prevent damage to components, then their use is described. Though it might be possible to borrow the tool, such work may have to be entrusted to the official agent.

To avoid labour costs a garage will often give a cheaper repair by fitting a reconditioned assembly. The home mechanic can be helped by this book to diagnose the fault and make a repair using only a minor spare part. The classic case is repairing a non charging dynamo by fitting new brushes.

The manufacturer's official workshop manuals are written for their trained staff, and so assume special knowledge; detail is left out. This book is written for the owner, and so goes into detail.

The book is divided into eleven Chapters. Each Chapter is divided into numbered sections which are headed in bold type between horizontal lines. Each section consists of serially numbered paragraphs.

There are two types of illustration: (1) Figures which are numbered according to Chapter and sequence of occurrence in that Chapter. (2) Photographs which have a reference number on their caption. All photographs apply to the Chapter in which they occur so that the reference figure pinpoints the pertinent section and paragraph number.

Procedures, once described in the text, are not normally repeated. If it is necessary to refer to another Chapter the reference will be given in Chapter number and section number thus: Chapter 1/16.

If it is considered necessary to refer to a particular paragraph in another Chapter the reference is eg, 'Chapter '1/5:5'. Cross references given without use of the word 'Chapter' apply to sections and/or paragraphs in the same Chapter, eg, 'see Section 8' means also 'in this Chapter'.

When the left or right side of a car is mentioned it is as if looking forward.

Great effort has been made to ensure that this book is complete and up to date. (Some owners of the new FIAT 126 may find it helpful). The manufacturers continually modify their cars, even in retrospect.

Whilst every care is taken to ensure that the information in this manual is correct no liability can be accepted by the authors or publishers for loss, damage or injury caused by any errors in or omissions from the information given.

Contents

Introduction to the Fiat 500		4
Vehicle identification and ordering spare parts		6
Routine maintenance		7
Lubrication chart		15
Recommended lubricants		16

Chapter		Page		Page
Chapter 1/Engine	Engine specifications	19	Cylinder head work	43
	Removing the engine	31	Engine reassembly	45
	Engine stripping	31	Engine replacement	50
	Component overhaul	40	Fault diagnosis	52
Chapter 2/Lubrication, cooling, heating and exhaust	Specifications	59	Overcooling	65
	General description	59	Heater	66
	Overheating	64	Exhaust	66
Chapter 3/Fuel system	Specifications	67	Idle adjustment	71
	General description	67	Petrol pump	72
	Carburettor	67	Fault finding	74
Chapter 4/Ignition system	Specifications	75	Ignition timing	77
	General description	75	Distributor	78
	Maintenance	76	Spark plugs	80
	Contact breaker adjustment	77	Fault finding	80
Chapter 5/Clutch	Specifications	83	Operating mechanism	84
	General description	84	Clutch cable	86
	Clutch assembly	84	Fault diagnosis	88
Chapter 6/Transmission	Specifications	89	Renovation	100
	General description	90	Reassembly	100
	Removal	90	Refitting	105
	Stripping	93	Fault finding	106
Chapter 7/Braking system	Specifications	107	Brake shoe replacement	111
	General description	107	Hydraulic system	111
	Bleeding the brakes	108	Handbrake	113
	Brake adjustment	108	Fault diagnosis	116
Chapter 8/Electrical system	Specifications	117	Starter motor	126
	General description	119	Windscreen wiper	129
	Battery	119	Lighting	130
	Generator	120	Instruments	130
	Control box	124	Fault finding	132
Chapter 9/Rear suspension, wheels and tyres	Specifications	137	Rear suspension	138
	General description	138	Shock absorbers	142
	Hub shaft	138	Wheels	142
	Hubs	138	Tyres	142
Chapter 10/Front suspension and steering	Specifications	143	Wheel alignment	148
	General description	144	Steering box	150
	Front hubs	144	Steering column	152
	Front suspension	144	Fault finding	154
Chapter 11/Bodywork	General description	155	Glass replacement	158
	Maintenance	155	Doors	158
	Repair	155	Trim	158
Metric conversion table				163
Index				165

Introduction to the Fiat 500

The FIAT 500 was introduced in 1957 as the Nuova Cinquencento, or New 500, to differentiate it from the old 500 "Topolino" or "Mouse". The Topolino had been in production from 1936; indeed its original drawings were started in 1916! It was very much a miniature car, conventional with four cylinder water cooled engine at the front driving the back axle.

The New 500, with 479 cc two cylinder air cooled engine at the rear was fundamentally an economy car designed most ably as such. Late in 1960 a larger engine of 499 cc was fitted, and the word "New" dropped from the model name, which was reclassified 500D. By 1965 modifications incorporated deserved reclassification again, to 500F. In 1969 this standard model was supplemented by a more fully equipped version, the 500L.

From 1960 to 1972 there was an estate car, sometimes called the Autobianchi Giardiniera. For most of that period there was also a van version of the station wagon, similar in specification except for the rear side windows.

For 1973 the engine was again enlarged, to 594 cc. This was at the time of the announcement of the larger bodied 126 FIAT. Both cars use the same engine, but the gearboxes are not shared the 500 keeping to its old one without synchromesh.

As well as the major changes that were marked by a reclassification of the model, numerous small modifications have been and still are being introduced.

Being one of the easiest cars to park, one of the cheapest to buy, and very economical on petrol, it is also cheap to repair. The economy of repair comes partly from the ease of doing most jobs, and also from the 500's robustness. There are not the weak points that are sadly more prevalent with other cars.

It will be a sad day when the 500 finally disappears for it has proven itself to be highly successful in this its original terms of reference - a cheap, robust town car. In addition it has also excelled for many people throughout Europe as a businesslike touring car as well as competition car.

FIAT 500 Station Wagon (1968)

FIAT New 500 (1958)

FIAT 500L (1972)

Vehicle identification and ordering spare parts

Numerous models of the FIAT 500 have been produced over more than a decade. There are, therefore, many different models which have been developed considerably from the first.

When identifying your own vehicle in order to purchase the correct spare part is essential to be specific. Always quote the chassis and engine numbers and state the year of manufacture, the model type and whether left or right hand drive.

FIAT franchised agents carry the largest stock of 'genuine' spare parts and certainly have a better knowledge than other garages. Pattern parts tend not to have the same quality and often do not fit properly and can be detrimental to performance and economy. If it is not possible to go direct to a FIAT agency, order through your local garage, still quoting the same vehicle numbers.

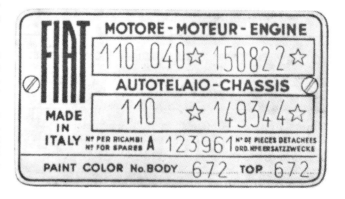

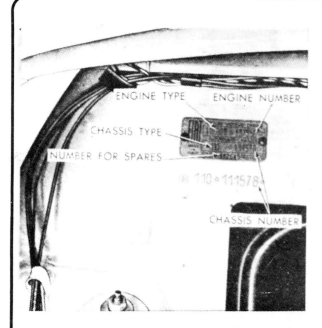

Identification plate giving engine and chassis number, and the spare part ordering number. Below the plate is the chassis number stamped on the body

Engine number on the crankcase

Routine maintenance

Introduction
1 In the paragraphs that follow are detailed the routine servicing that should be done on the car. This work has two important functions. First is that of doing adjustments and lubrication to ensure the least wear and most efficient function. But the second gain from maintenance, could almost be more important. By looking your car over, on top and underneath, you have the opportunity to check that all is in order.
2 Every component should be looked at, your gaze working systematically over the whole car. Dirt cracking near a nut or a flange can indicate something loose. Leaks will show. Electric cables rubbing, rust appearing though the paint underneath, will also be found before they bring on a failure on the road, or a more expensive repair if not tackled quickly. Even a little FIAT 500 is lethal if unsafe.
3 The tasks to be done on the car are in general those recommended by the maker. We have also put in some additional ones. For someone getting his servicing done at a garage it may be more cost effective to accept component replacement after a somewhat short life in order to avoid maintenance costs. For the home mechanic this is not so. The manufacturers must detail the work to be done as a careful balance of such factors. Leaving it too long gives risk of defects occuring between service checks. Making intervals too frequent tempts owners into disrespect of their advice, to leave work undone disastrously long.
4 When you are checking the car, if something looks wrong look it up in the appropriate Chapter. If something seems to be working badly look in the fault finding section.
5 Always road test after a repair, and inspect the work after it, and check nuts etc for tightness. Check again after about 150 miles.

Tools
1 The most useful type of spanner is a "combination spanner". This has one end open jaw the other a ring of the same size. In any case it is difficult to buy double ended spanners in metric sizes with the second end having the next size that you want; they are usually only 1 mm different, and you do not need all of them.
You need two spanners of each size so that one is on a nut whilst the other holds the bolt.
2 The sizes you need are:-

8, *10*, 12, *13*, 14, *17*, 22, 24 and 27 mm.
The most frequently used ones are in italic. If you already have some inch spanners there is in several cases a good match.

2 BA for 8 mm
½ AF for 13 mm but too tight in some cases
¼ Whit for 13 mm but very loose
9/16 AF for 14 mm but very loose
11/16 AF for 17 mm but a bit loose
¾ AF for 19 mm but a bit loose
7/8 AF for 22 mm but a bit loose
11/16 AF for 27 mm Good fit

3 You will need a set of feeler gauges. Preferably these should be metric. In many cases we quote the inch equivalent of dimensions, and conversion tables are at the back of the book. But the car is metric, and errors will be avoided by working metric. In the longterm metric equipment will be a good investment.
4 You will see we specify tightening torques for nuts. This needs an expensive torque wrench. Many people get on well without them. Contrariwise many others are plagued by things falling off or leaking from being too loose, whilst others suffer broken bolts, stripped threads, or warped cylinder heads, because of overtightening.
5 Torque wrenches use the socket of normal socket spanner sets. Sockets, with extensions and ratchet handles, are a boon. In the meantime you will need box spanners for such things as cylinder head nuts, and the spark plugs. They are thinner than sockets in small sizes, and will go where the latter cannot, so will always be useful even if later you plan to get sockets.
6 Screwdrivers should have large handles for a good grip. You need a large ordinary one, a little electrical one, and a medium cross-headed one. Do not buy one handle with interchangeable heads. The large screwdriver must have a tough handle that will take hitting with a hammer when you misuse it as a chisel.
7 You can use an adjustable spanner and a self grip or pipe wrench of the Mole or Stillsons type.
8 With these tools you will get by. Do not do as much as we do in the photos; whenever possible use a ring spanner (or socket). We often show an open jaw spanner in use. This is to put its size marking in the photo.
9 If you undertake major dismantling of the engine or transmission you will need a drift. This is a steel rod or "tommy bar" about 3/8 inch diameter, and made of tough steel to stand hammering. Do not use a brass drift lest little chips are cut off and get into some component and ruin it. You will need a "ball/pane" hammer, fairly heavy too because it is easier to use gently, than a light one hard.
10 Files are soon needed. Four makes a good selection:

6 inch half round smooth
8 inch flat second cut
8 inch round second cut
10 inch half round bastard

11 You need a good, firm, hydraulic jack. A trolley jack is of major value for removing the engine. If you do ever get one, it must be in addition to and cannot replace the simple jack, which is needed too for smaller jobs.

Maintenance tasks

Weekly or 300 Miles (500 km) if sooner: or before a long journey.

1 Check tyre pressures:

Front	Rear (light load)	Rear (full load)	
18 (17)	23 (27)	27 (30)	lbf/in^2
1.3 (1.2)	1.6 (1.9)	1.9 (2.2)	kgf/cm^2

Figures in brackets for station wagon.
For radial tyres the front should only be 16 lbf/in2 (1.1 kgf/cm2.)
2 Check oil level in engine sump.
3 Check the level of brake hydraulic fluid in the reservoir under the front bonnet.
4 Check all lights are working. A convenient way to do the brake lights is to reverse near something shiny.

Monthly: or 1500 miles (2,500 km) if sooner.

Do all the weekly/300 miles tasks and in addition:
1 Check battery. Top up electrolyte to just above the plate separators with distilled water. Remove any corrosion and smear battery posts and terminal fittings with vaseline. If corrosion recurs take the terminals off and try to be more thorough.
2 Top up the windscreen washer reservoir with a mixture of water and detergent. (Mild household washing up liquid).

Every 3,000 Miles (5,000 km)

Do all the more frequent tasks and in addition:-
1 Check the car underneath. Look at the rubber dirt excluders on the steering ball joints. Check the flexible hydraulic pipes for the brakes (one at each front wheel, one just inside each rear suspension forward pivot), for rubbing or leaks.
2 Check under the car for oil leaks (or new oil leaks if minor seepage already occurs).
3 Lubricate the grease nippled on the king pins. Jack up the car to get the weight off the front suspension. Clean the nipples (one each side). Pump in grease till it exudes clean at the ends of the king pin. Wipe off excess grease (a little left around helps keep out wet).
4 Check the fan belt tension. It should be possible to press it down ½ inch (1 cm) using one finger and a hard pressure. To tighten, undo the three nuts clamping the two parts of the dynamo pulley, and moce a spacer from between the two halves to the outside. Do not overtighten or the dynamo bearings will be overloaded. Excessive slackness wears the belt by slip. New belts stretch, and require checking two or three times in their first few hundred miles.

Every 6,000 Miles (10,000 km) or 6 Months if sooner:

Do all the more frequent tasks, and in addition:
1 **Engine oil change.** Drain the oil from the plug on the right of the sump when the car is hot, (17 mm spanner), into a pan that will have plenty of room with the 4½ pints in it, to prevent spillage. Allow the oil to drip for at least ten minutes. Clean the drain plug, and check its washer. Refit the plug, and refill the sump with 4 pints. Check the level on the dipstick and add a little more if required. The oil should be changed at 3,000 miles if the car is used mostly for town work, particularly in winter, or in dusty conditions. Use a top quality multigrade oil (Castrol GTX), without any other additives. If you have just bought an old car and the oil is very black flush out first with flushing oil. When draining this jack up the left of the car once most of the oil is out to make any dregs flow out better.
2 **Air cleaner** - renew the element.

3 **Spark plugs** - fit clean plugs. With the gap set to .025 in (.6 mm) plugs are Marelli CW 6N, Champion L87Y (or L7) or Bosch W225 T 1. Have two sets, the ones taken out being sandblasted by the local garage ready for the next change. Plugs should be replaced after the second time of use. The porcelain plug insulators. and the leads, must be kept clean to ensure easy starting.
4 **Distributor**
Undo the two clips and remove the cap. Pull off the rotor arm. Check that the points are not contaminated with oil or dirt. Check their gap. This should be .019 to .021 inch (.47 - .53 mm). This may be almost impossible to measure other than by eye, as a lump will have burned onto one contact, preventing insertion of a feeler. If in doubt take out the contacts and clean them by rubbing off the lump on an oil stone (or wet doorstep). See Chapter 4:5 for details. Lubricate the wick on the spindle normally covered by the rotor arm with engine oil, and oil the lubricator just outside the distributor. Add a drop of oil to the contact breaker pivot and to the centrifugal advance mechanism beneath the contact breaker. Clean the inside and outside of the distributor cap, and the rotor arm.
5 **Ignition timing**
Check the timing. It will definitely need adjustment if the contact breaker has been adjusted. The timing should be 10° before top dead centre, TDC. There is a timing mark for TDC on the timing chain cover on the rear of the engine. 10° BTDC can be established by measuring 13 mm (.51 inch) from the TDC mark. See Chapter 4:6 for details.
6 **Transmission oil**
Check the oil level in the transmission. The car should be on level ground. The level plug (13 mm spanner) is a small one, on the right side of the casing, just behind the axle dirt excluder. The level should be up to the bottom of the thread holes. Fill with Castrol Hypoy 90, which can be bought in convenient plastic 'flexitops' complete with spout. Before removing the plug clean carefully around and above it. Wipe again after the plug has been loosened a turn.
7 **Oil can lubrication**
Put a few drops of engine oil from an oil can on all parts such as door hinges, their catches (the 500L has little plastic plugs in the holes), control pedal pivots, carburettor controls and cables, and the front and rear bonnet catches and hinges. Lubricate the handbrake cable fixing pins at the back of the rear brakes.
8 **Clutch**
Check the free play between the pedal and the clutch. The pedal should move for ½ - ¾ inch (15 - 20 mm) for 110F cars with diaphragm clutches, and 1½ ins. (35 - 40 mm) for all earlier ones, against only the light load of its pull-off spring before taking up the clutch pressure. If necessary adjust where the cable is attached to the clutch withdrawal lever on the transmission casing above the left drive shaft.
9 **Tappets**
Check the valve clearance. It should be .006 in (.15 mm) cold. The engine can be turned over using a 10 mm spanner on the oil filter bolts. Adjust the tappets of one cylinder when those of the other are just at "change over": the exhaust will have just shut and the inlet be about to open; both will have no clearance. Slide in the feeler gauge on one of the tappets being adjusted. It should slide in readily, but resistance be felt. Trying to twist round the push rod gives a good guide. A feeler the next size smaller should slide in decidedly freely, and one a size too large should be very stiff, having to be forced in, and then it is impossible to rotate the push rod. Repeat for the other valve of that cylinder, then do the second cylinder after turning the engine over one revolution. After that do the first again just to check your own measurement has not been at fault.
10 **Fuel system**
Clean the filters. There is one in the pump, reached by undoing the screw holding the domed cover on top. On some cars it is in the inlet union. See Chapter 3. In the carburettor there is one under a large nut on the float chamber, near the fuel pipe union. If the car is new, so the tank clean, and always

RM1 The dipstick and oil filler

RM2 The hydraulic brake fluid reservoir

RM3 King pin grease nipple (one each side)

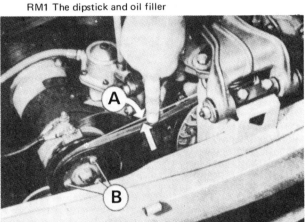
RM4 (A) The fan belt should yield ½ in (B) Generator pulley bolts

RM5 The engine sump drain plug

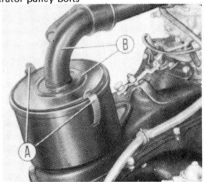
RM6 Sedan air cleaner (A) Body top retaining clips (B) Carburettor intake and body top

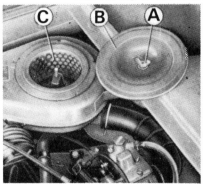
RM7 Station wagon air cleaner - front right cover of engine
Undo the butterfly nut (A), lift off the lid (B), to get at the element (C).

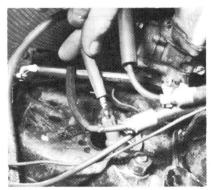
RM8a Plug change: first pull off the lead

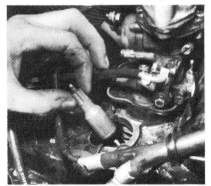
RM8b Unscrew terminal spacer. (Station wagon has dirt excluder - lead goes on plug)

RM8c You must use self gripping plug spanner. Plug falls into air cowling if dropped

refuelled under clean conditions you may find very little dirt on these filters. If this is so, this task may be done at double the interval. Therefore in the summary it is under 12,000 miles task.

11 Brakes

Check the brakes for wear. On the saloon/sedan they are self adjusting, but the linings could be worn down too thin. Remove the brake drums and look inside. See Chapter 7 for details. The minimum safe thickness of lining left on the shoe is 0.59 ins (1.5 mm). Clean out all dust from the drum and shoe assemblies. Check the hydraulic operating cylinder for fluid leaks. Dampness around the dirt excluder is acceptable, but if a leak is bad enough for fluid to seep further then replacements are necessary straightaway. If the brake shoes are worn to the minimum thickness replace them immediately. Check the drums for scoring, or cracks. On the station wagon adjust both brake shoes of all four wheels. (See Chapter 7). On all cars check the travel of the handbrake. If it moves more than four clicks readjust it. Tighten the cable ends at each rear wheel. There is a balancing mechanism, but it will help the cable life in its sharp bend at the lever if both are done. With the car jacked up and both rear wheels removed, pull the lever up two clicks, on its ratchet. Slacken the adjuster locknuts. Tighten the adjusters whilst rotating the brake drum to feel for the pull of the brakes. Tighten till the drum can only just be turned by hand. Later, road test the car. Drive carefully so the brakes are not used. Stop when going up hill by using the gears. Check the temperature of the brake drums by hand behind the wheel. All should be cool after such driving; and the back should be as cool as the front. If they are not it indicates the handbrake is too tight.

12 Steering

Check for wear. Get an assistant to move the steering wheel to and fro just enough to move both front wheels a little. Watch all the steering knuckles on both ends of all three track rods: there should be no visible free play. The idler arm on the opposite side of the car to the steering box should be firm, and not move up and down, though it is slightly resilient. No movement other than that for moving the steering should be visible on the steering arm coming from the steering box. Now reduce the turning of the steering wheel to see how far it can be moved without the wheels moving. This will not be very definite, the various clearances in the steering box adding up to a slight springy feeling. The lost motion should not exceed ½ inch at the steering wheel rim. The steering wheel should have no movement at all when pulled up and down axially (in line with the column down to the box). A little slop of the column in its top bearing is permissible. See Chapter 10 for detailed checks and adjustment. Refill the steering box with SAE 90EP oil to the level of the plug (item 27 in Fig.10.6).

13 Tyres

Measure the wear. The legal limit is 1 mm of tread left. The safety limit under wet conditions gets low at 2 mm. Note for any uneven wear. Keep a record of the tread remaining. Then it is possible to see which tyre is wearing the fastest. We do not recommend changing the tyres round frequently. See Chapter 10. Examine the tyres for cuts or bulges, particularly in the sides. Do not run on such tyres. Not only is it illegal, but dangerous too.

14 Engine test

Start up, and warm up the engine to check it at idle. The idle should be slow and smooth. The speed should be such that the red charging warning light just comes on. Adjust the speed by the throttle stop screw and the smoothness by the mixture screw (see Chapter 3:4). If the engine was in need of attention to such things as tappets or ignition then the idle is likely to need readjusting too. Whilst listening to the engine listen for odd noises, and exhaust leaks in particular. (If suspicious see Chapter 2). Get used to the noises of your engine idling, when running a bit faster, and when you "blip" the throttle. Listen to the cars of other peoples too. Then you will know what are correct sorts of noises, and what defects. Check all the bolts on the exhaust for tightness, and those holding down the carburettor.

15 Bodywork

Under this heading is work of different type to previous tasks. This task can most easily be done at some seperately, at some different time to the mechanical jobs that are more of a mileage type; the bodywork checks need doing more by time as a 6 monthly task than by mileage, and as such are best done in spring and autumn. Hose the car down thoroughly underneath, preferably after a journey on wet roads so the mud has had time to soak until soft. Pay particular attention to corners and ledges. Allow the car to dry off; this may take over 24 hours; it is helped if the car can be parked in a good drying wind. Jack the car up high, by going up in stages, using planks under the wheels. Make the car firm on additional blocks. Take off the wheels one by one to deal with each wing. For security only one wheel off at a time. The car must be examined for peeling underneath paint. Sometimes the paint can be present, but flaking, with damp trapped under it, rusting fast. Remains of mud may have to be scraped off to get down to the surface. Only scrape where you will be able to repaint, as the slightest cut of the scraper will allow damp in. Spray into all corners with a rust inhibitor aerosol that is also a thick preservative such as 'Supertrol 001' or 'Di-Nitrol 3B'. Then paint with 'Bronze Adup' thick protective paint. This paint is compatible with the two aerosols. Pay particular attention to welded joints, as the stress put in these at the time of welding promotes corrosion. Cover more thickly the parts where the wheels throw up gravel. Do the same to the underbody between the front and rear axles. Squirt the aerosol into all hollow sections. Squirt it up through the door drain holes, at the same time checking that these are clear. Lower the car. Check the paintwork on the body for little chips of paint. Touch up, first with primer such as 'Belco', then the car's matching touch-up paint. The 500 is quite good against rust. The worst places are the scoop at the front in which sits the spare wheel, both inside and outside, and a little transverse reinforcing section crosswise under the floor level with the front seats.

Every 12,000 Miles (20,000 km) or annually if sooner.

Do all the more frequent tasks, and in addition:

1 Change the oil in the transmission. When the car is hot after a journey drain the oil from the plug in the bottom of the casing (17 mm spanner). Clean the plug before taking it out. The container should have room for 2 pints without risk of overflowing. Clean the plug before putting it back. Clean the level plug, then take it out and refill with SAE 90 EP oil. About 2 pints is needed. Use oil such as Castrol Hypoy 90, which comes in plastic containers with spouts. Fill till it just reaches the level of the plug hole.

2 Front hubs

It is a task at this mileage to check the endfloat in the front hubs. This will be done anyway whilst examining the front brakes, as part of the 6,000 mile task.

3 Toe-in

Check the alignment of the front wheels. They should toe-in between 0 and 2 mm (about 1/16 inch) (estate car 0 - 1 mm). This is a difficult job. Get a garage to do it, or see Chapter 10. Incorrect toe-in is a frequent source of rapid front tyre wear. Again, see Chapter 10.

4 Carburettor

The float chamber and passages should be cleaned out. Clean the outside of the carburettor. Take off the air pipe from the cleaner and the fuel pipe from the pump. Undo the screws holding the top of the carburettor. Carefully lift off the top. With this will come the float and needle valve assembly. Remove the main jet from outside the carburettor (see Chapter 3:4). With a fluff-free cloth wipe out any dirt in the float chamber. Take care not to get any dirt in the float chamber. Take care not to get any dirt on the other jets which are exposed. Pour a little petrol into the float chamber so that it swills dirt out through the main jet hole. Blow through then replace the jet, and reassemble the carburettor. Note cleanliness is essential for this operation, otherwise you may be worse off than before.

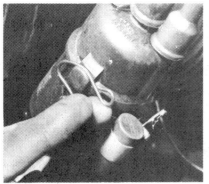
RM9a Unclip the distributor

RM9b Pull off the rotor arm and then....

RM9c ....check the contact breaker points gap

RM10a To tighten (or change) fan belt undo the 3 nuts clamping pulley halves

RM10b Move packing pieces one at a time to the outside. Narrower pulley squeezes out the belt

RM10c To adjust the tappets undo the 2 nuts and lift off rocker cover

RM10d Turn over engine till valves of other cylinder are at "change over", insert feeler. Hold adjuster whilst unlocking nut

RM10e Take off domed cover

RM10f Clean filter. Later pumps do not have one

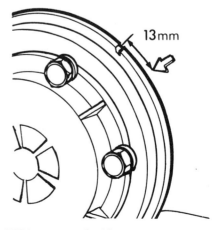
RM11a The timing mark on the engine gives TDC. The required 10° BTDC is measured as 13 mm round the pulley

5 Engine breather

Clean the filter that rests inside the oil filler on the rocker-box. This can be lifted out once the filler cap has been removed. Check the condition of the fragile valve on the filler cap. If the valve does not seat properly, or if a bad step is worn on it replace it. See Chapter 1 for defects.

Prise off the breather pipe from its connection to the air intake near the air cleaner. With a piece of wire mesh inside for the flame trap; it looks like a bottle brush. Clean the flame trap and the filler filter by swilling them in a clean bowl of petrol. These components are not fitted in the breathers of cars made before 1965. On station wagons the flame trap is at the oil filler end of the breather pipe to the air cleaner.

6 Engine oil filter

Remove the six bolts (10 mm) from the centrifugal oil filter cover. The fan belt drive pulley houses this filter. Put a piece of rag underneath to catch the oil that will pour out. Carefully take off the cover. No not prise with a sharp instrument or the soft metal face will be marked and may leak. Take out the rubber sealing ring. Scrape out the dirt collected at the outside of the filter, from both the half still on the engine, and the cover. Take care no dirt gets into the oil passages of the engine that lead to and from the filter (see photographs in Chapter 1), near the centre of the filter. Reassemble using a new sealing ring. Tighten the bolts gradually, evenly, and diagonally.

7 Check tightness

Put a spanner on and check for tightness the nuts/bolts: Holding the rear panel across the engine compartment, and the engine mountings on it. Under the car the bolts for the rear suspension to the body, the bracket holding the transmission to the bottom of the car, the steering box and the idler lever to the front of the body. Check the door hinges and locks and striker plates.

8 In the middle of winter give the battery a long slow charge.

Every 18,000 Miles (30,000 km)

Do all the more frequent tasks except those for 12,000 miles (unless 36,000 or 72,000 miles run) and in addition:

1 Dynamo check

Remove the dynamo. See Chapter 8/9 - 11. Clean the commutator and renew the brushes. Re-lubricate the bearings. Refit using a new fan belt.

2 Starter motor check

FIAT recommend the replacement of starter brushes at 18,000 miles. This is for a car used about town, which will mean short journeys with frequent use of the starter. Under these conditions, or if the car has been bought recently second hand and its history is unknown then the service should be done. Under good conditions this task can be left to double the mileage, or three years (for lubrication reasons). The work to be done is to fit new brushes and lubricate the pinion free wheel. See Chapter 8.

3 Wheel bearings

Repack both front and rear hubs with grease. Readjust the front hub endfloat on reassembly. Check that of the rear hub, which depends on a resilient bearing spacer: See Chapter 9/3:7 for the rear, and Chapter 10 for the front.

Other aspects of Routine Maintenance

1 Jacking up

Always chock a wheel on the opposite side in front and behind. The car's own jack has to be able to work when the car is very low with a flat tyre, so it goes in a socket on the side, taking up both wheels on that side. Using a small jack at wheel is more secure when work has to be done. At the front it should be put under the spring, as close as it will go on the flat of the leaf to the eye at the end. At the back put a jack under the suspension arm as near as possible to the wheel, in the centre of the large reinforced area beneath the drive shaft coupling, but with a wooden plank between jack and arm to spread the load. There are jacking points suitable for a trolley jack reinforcing the centre of the front and the rear panels; put wood blocks between them and the jack head. The front one is also the towing eye. Never put a jack under the bodywork or the thin sheet steel will buckle.

2 Wheel nuts

These should be cleaned and lightly smeared with grease as necessary during work, to keep them moving easily. If the nuts are stubborn to undo due to dirt and overtightening, it may be necessary to hold them by lowering the jack till the wheel rubs on the ground. Normally if the wheel brace is used across the hub centre a foot or knee held against the tyre will prevent the wheel from turning, and so save the wheels and nuts from wear if the nuts are slackened with weight on the wheel. After replacing a wheel make a point later of rechecking the nuts again for tightness.

3 Safety

Whenever working, even partially, under the car, put an extra, strong, box or baulk of timber underneath onto which the car will fall rather than onto you.

4 Cleanliness

We had two FIAT 500's on which we worked for this book. The one that was filthy made life unpleasant, and until we cleaned it gave risk of contaminating new components. The cleanliness of the other made work more pleasant. Whenever you do any work allow time for cleaning. When something is in pieces, components removed improve access to other areas and give an opportunity for a thorough clean. This cleanliness will allow you to cope with a crisis on the road without getting yourself dirty. During bigger jobs when you expect a bit of dirt it is less extreme. When something is taken to pieces there is less risk of ruinous grit getting inside. The act of cleaning focusses your attention on parts, and you are then more likely to spot trouble. Dirt on the ignition parts is a common cause of poor starting. Large areas such as the engine compartment bulkheads and the cowling should be brushed thoroughly with detergent like Gunk, allowed to soak, then very carefully hosed down. Water in the wrong places, particularly the carburettor or electrical components will do more harm than dirt. Detailed cleaning can be done with petrol/paraffin mix and an old paint brush.

5 Waste disposal

Old oil and cleaning paraffin must be destroyed. It makes a good base for a bonfire, but is dangerous. Never have an open container near a naked flame. Pour the old oil where it cannot run uncontrolled, before you light it. Light it by making a "fuse" of newspaper. By buying your new engine oil in one gallon cans have these for storage of the old oil. The old oil is not "household rubbish", so should not go in the dustbin. Most councils have area collection points where other types of rubbish can be put. (Under the Civic Amenities Act).

6 Long journeys

Before taking the car on long journies, particularly such trips as long continental holidays, do in advance many of the maintenance tasks that would normally not be due before going. In the first instance do jobs that would be due soon anyway, but also those that would come up well into your trip but before you get back. Also do all the checks of other tasks as a kind of insurance. Carry on the car some copper wire, plastic insulation tape, plastic petrol pipe, Hermetite golden gasket compound and repair material such as "Plastic Padding Hard". Carry a spare fan belt, and some spare bulbs. About 3 foot of electric cable, and some odd metric nuts and bolts should complete your car's 'first aid kit'. Also carry a human first aid kit; some plasters, germolene ointment, etc, in case you get minor cuts.

7 On purchase

If you have bought your 500 brand new you will have the makers instructions for the early special checks on a new car. If you have just bought a second hand car then our advice is to assume it has not been looked after properly, and so do all the checks, lubrication and other tasks on that basis, assuming all mileage and time tasks are overdue.

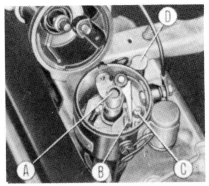

RM1 lb Use engine oil on wick (A) and oil hole (D). Also oil contact pivot with just one drop and automatic advance mechanism underneath. Contacts (B) are clamped by screw (C)

RM12 Transmission level and drain plugs

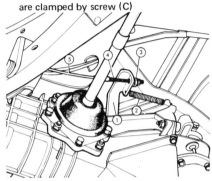

RM13a Adjusting clutch to give ½ in free play at pedal. Lever (1), pull-off spring (2) are pulled by cable (5) with adjuster and locknut (3) on threaded end (4)

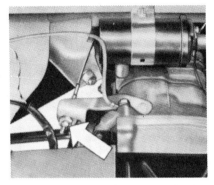

RM13b On station wagon clutch adjuster can be reached from above

RM14 Adjusting tappets. See Chapter 1

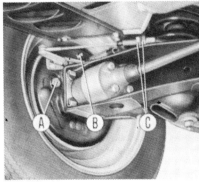

RM15 Brake adjusters on station wagon

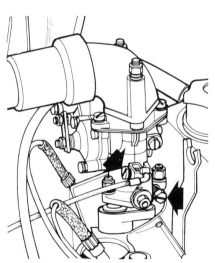

RM16 Two screws controlling the idle on Sedan's carburettor. See Chapter 3/4

RM17 Throttle stop (A) and idle mixture screw (B) on station wagon's carburettor

RM18 Oil filter is centrifugal. Slacken bolts (10 mm) evenly, and diagonally

Maintenance Summary

Weekly: or 300 Miles (500 km)

1. Tyre pressures
2. Engine oil level
3. Brake fluid level
4. Check all lights

Monthly: or 1500 Miles (2,500 km)

All weekly tasks and:-
1. Battery electrolyte level
2. Windscreen washer reservoir

3,000 Miles (5,000 km)

Weekly and monthly tasks and:-
1. Check underneath (loose or leaks)
2. Lubricate steering king pins
3. Check fan belt tension

6,000 Miles (10,000 km): or 6 months

Weekly/Monthly/3000 Miles tasks and:-
1. Engine oil change
2. Engine air cleaner new element
3. Clean spark plugs
4. Distributor lubrication
5. Contact breaker check
6. Ignition timing check
7. Transmission oil level check
8. Oil can lubrication hinges/controls
9. Clutch pedal free play
10. Tappet clearance
11. Check brake wear
12. Check steering wear. Refill steering box with oil
13. Check tyre wear
14. Check engine idle
15. Bodywork rust prevention

12 000 Miles (20,000 km): or annually

All more frequent tasks and:-
1. Change the transmission oil
2. Check front wheel toe-in
3. Clean engine breather
4. Clean engine oil centrifugal filter
5. Clean fuel pump and carburettor filters
6. Check bolts for tightness
7. Recharge the battery in mid winter

18,000 Miles (30,000 km)

All more frequent tasks except 12,000 and:-
1. Dynamo brushes and lubrication
2. New fan belt
3. Starter motor check
4. Repack wheel bearings

Read through the previous section 'Maintenance Tasks' before starting.

Lubrication chart

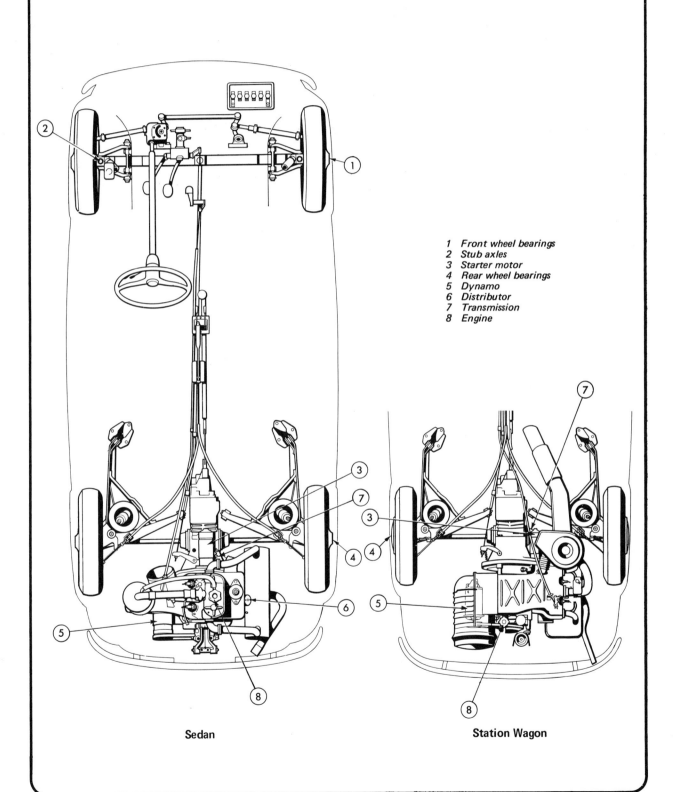

1 Front wheel bearings
2 Stub axles
3 Starter motor
4 Rear wheel bearings
5 Dynamo
6 Distributor
7 Transmission
8 Engine

Sedan

Station Wagon

Quick glance dimensions and capacities

Dimensions	Fiat 500 De-luxe	Fiat 500 D.	Fiat 500 Station Wagon
Length, overall, with bumpers	119.09 in (3.025 m)	116.93 in (2.970 m)	125.39 in (3.185 m)
Width, overall	51.97 in (1.320 m)	51.97 in (1.320 m)	51.97 in (1.320 m)
Height, unladen	52.56 in (1.335 m)	52.56 in (1.335 m)	53.30 in (1.354 m)
Wheelbase	72.44 in (1.840 m)	72.44 in (1.840 m)	76.38 in (1.940 m)
Tread, front	44.13 in (1.121 m)	44.13 in (1.121 m)	44.13 in (1.121 m)
Tread, rear	44.68 in (1.135 m)	44.68 in (1.135 m)	44.53 in (1.131 m)
Ground clearance	4.92 in (0.125 m)	4.92 in (0.125 m)	5.31 in (0.135 m)
Turning radius	14 ft 1 in (4.300 m)	14 ft 1 in (4.300 m)	14 ft 1 in (4.300 m)
Weights			
Kerb weight	1168 lb (530 kg)	1124 lb (510 kg)	1234 lb (560 kg)
Distribution - front rear (unladen)	43/57%	42/58%	39/61%
Payload	4 persons plus 88 lb	4 persons plus 88 lb	4 persons plus 88 lb or 1 person plus 551 lb

Capacities	Imperial	Metric	American
Fuel tank	4½ gals	21 litres	5½ US gals
Engine sump refill	4¼ pints	2½ litres	5¼ US pints
(Early cars)	3 pints	1¾ litres	3½ US pints
Transmission oil	1.9 pints	1.1 litres	2.3 US pints

Tyre pressures	Sedan (Crossply)	Sedan (Radial)	Station wagon
Front (lbs/in) (kg/cm)	18 (1.3)	16 (1.1)	17 (1.2)
Rear	23 (1.6)	23 (1.6)	27 (1.9)
Rear fully laden	27 (1.9)	27 (1.9)	30 (2.2)

Recommended lubricants

	TYPE OF LUBRICANT	CORRECT CASTROL GRADE
ENGINE	Monograde oil applicable	Castrol Multigrade Oil
Air temperature constantly in the range of:		
above 86°F) above 30°C)	SAE 40	20W - 50 Castrol GTX
32° - 86°F) 0° - 30°C)	SAE 30	20W - 50 Castrol GTX
5° - 32°F) −15° - 0° C)	SAE 20E	10W - 40 Castrolite
Below 5°F) Below −15°C)	SAE 10W	10W - 40 Castrolite
TRANSMISSION	SAE 90 EP	Castrol Hypoy

	FIAT GREASES	CASTROL PRODUCT
King pins	JOTA 1)	High melting point lithium-based grease Castrol LM Grease
Hubs	MR3)	
Dynamo	MR 3)	Molybdenum di-sulphide based grease Castrol MS3 Grease
Starter motor	MR 2)	
Brake fluid	FIAT Azzura	Fluid to Specification SAE J1703b Castrol Girling Brake Fluid

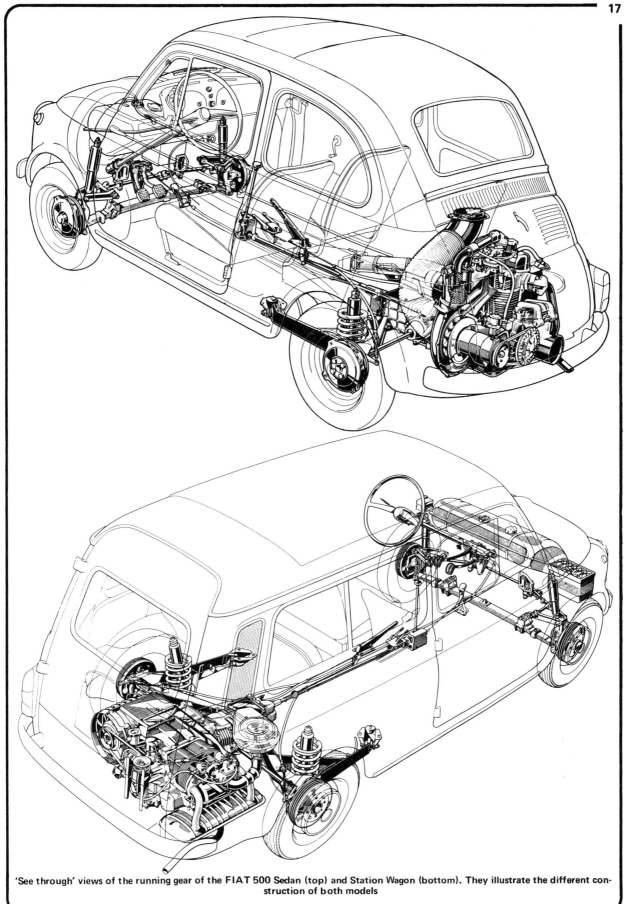

'See through' views of the running gear of the FIAT 500 Sedan (top) and Station Wagon (bottom). They illustrate the different construction of both models

Castrol GRADES

Castrol Engine Oils

Castrol GTX
An ultra high performance SAE 20W/50 motor oil which exceeds the latest API MS requirements and manufacturers' specifications. Castrol GTX with liquid tungsten† generously protects engines at the extreme limits of performance, and combines both good cold starting with oil consumption control. Approved by leading car makers.

Castrol XL 20/50
Contains liquid tungsten†; well suited to the majority of conditions giving good oil consumption control in both new and old cars.

Castrolite (Multi-grade)
This is the lightest multi-grade oil of the Castrol motor oil family containing liquid tungsten† It is best suited to ensure easy winter starting and for those car models whose manufacturers specify lighter weight oils.

Castrol Grand Prix
An SAE 50 engine oil for use where a heavy, full-bodied lubricant is required.

Castrol Two-Stroke-Four
A premium SAE 30 motor oil possessing good detergency characteristics and corrosion inhibitors, coupled with low ash forming tendency and excellent anti-scuff properties. It is suitable for all two-stroke motor-cycles, and for two-stroke and small four-stroke horticultural machines.

Castrol CR (Multi-grade)
A high quality engine oil of the SAE-20W/30 multi-grade type, suited to mixed fleet operations.

Castrol CRI 10, 20, 30
Primarily for diesel engines, a range of heavily fortified, fully detergent oils, covering the requirements of DEF 2101-D and Supplement 1 specifications.

Castrol CRB 20, 30
Primarily for diesel engines, heavily fortified, fully detergent oils, covering the requirements of MIL-L-2104B.

Castrol R 40
Primarily designed and developed for highly stressed racing engines. Castrol 'R' should not be mixed with any other oil nor with any grade of Castrol.

†*Liquid Tungsten is an oil soluble long chain tertiary alkyl primary amine tungstate covered by British Patent No. 882,295.*

Castrol Gear Oils

Castrol Hypoy (90 EP)
A light-bodied powerful extreme pressure gear oil for use in hypoid rear axles and in some gearboxes.

Castrol Hypoy Light (80 EP)
A very light-bodied powerful extreme pressure gear oil for use in hypoid rear axles in cold climates and in some gearboxes.

Castrol Hypoy B (90 EP)
A light-bodied powerful extreme pressure gear oil that complies with the requirements of the MIL-L-2105B specification, for use in certain gearboxes and rear axles.

Castrol Hi-Press (140 EP)
A heavy-bodied extreme pressure gear oil for use in spiral bevel rear axles and some gearboxes.

Castrol ST (90)
A light-bodied gear oil with fortifying additives

Castrol D (140)
A heavy full-bodied gear oil with fortifying additives.

Castrol Thio-Hypoy FD (90 EP)
A light-bodied powerful extreme pressure gear oil. This is a special oil for running-in certain hypoid gears.

Automatic Transmission Fluids

Castrol TQF
(Automatic Transmission Fluid)
Approved for use in all Borg-Warner Automatic Transmission Units. Castrol TQF also meets Ford specification M2C 33F.

Castrol TQ Dexron®
(Automatic Transmission Fluid)
Complies with the requirements of Dexron® Automatic Transmission Fluids as laid down by General Motors Corporation.

Castrol Greases

Castrol LM
A multi-purpose high melting point lithium based grease approved for most automotive applications including chassis and wheel bearing lubrication.

Castrol MS3
A high melting point lithium based grease containing molybdenum disulphide.

Castrol BNS
A high melting point grease for use where recommended by certain manufacturers in front wheel bearings when disc brakes are fitted.

Castrol CL
A semi-fluid calcium based grease, which is both waterproof and adhesive, intended for chassis lubrication.

Castrol Medium
A medium consistency calcium based grease.

Castrol Heavy
A heavy consistency calcium based grease.

Castrol PH
A white grease for plunger housings and other moving parts on brake mechanisms. *It must NOT be allowed to come into contact with brake fluid when applied to the moving parts of hydraulic brakes.*

Castrol Graphited Grease
A graphited grease for the lubrication of transmission chains.

Castrol Under-Water Grease
A grease for the under-water gears of outboard motors.

Anti-Freeze

Castrol Anti-Freeze
Contains anti-corrosion additives with ethylene glycol. Recommended for the cooling systems of all petrol and diesel engines.

Speciality Products

Castrol Girling Damper Oil Thin
The oil for Girling piston type hydraulic dampers.

Castrol Shockol
A light viscosity oil for use in some piston type shock absorbers and in some hydraulic systems employing synthetic rubber seals. It must not be used in braking systems.

Castrol Penetrating Oil
A leaf spring lubricant possessing a high degree of penetration and providing protection against rust.

Castrol Solvent Flushing Oil
A light-bodied solvent oil, designed for flushing engines, rear axles, gearboxes and gearcasings.

Castrollo
An upper cylinder lubricant for use in the proportion of 1 fluid ounce to two gallons of fuel.

Everyman Oil
A light-bodied machine oil containing anti-corrosion additives for both general use and cycle lubrication.

Chapter 1 Engine

Contents

General description ... 1	**PART C - REASSEMBLY**
Planning work ... 2	Preparation for reassembly ... 23
	Crankshaft and main bearings - assembly ... 24
PART A - DISMANTLING	Connecting rods, pistons, cylinders, big-ends - assembly ... 25
Removing the cylinder head in situ ... 3	Crankcase minor components - reassembly ... 26
Removing engine less transmission ... 4	Flywheel - refitting ... 27
Removing engine with transmission ... 5	Oil pump - reassembly ... 28
Prepatory stripping (including air cowling and dynamo) ... 6	Camshaft ... 29
Removing the head on the bench ... 7	Timing chain - reassembly ... 30
Stripping the centrifugal oil filter ... 8	Oil filter - reassembly ... 31
Timing chain removal ... 9	Cylinder head - replacement ... 32
Removing the flywheel ... 10	Air cowling and dynamo - refitting ... 33
Minor crankcase components ... 11	Replacing the engine - less transmission ... 34
Camshaft - removal ... 12	Replacing the engine - with transmission ... 35
Connecting rods and pistons ... 13	Final assembly and starting up ... 36
Crankshaft and main bearings - removal ... 14	
Stripping the oil pump ... 15	**PART D - ADDITIONAL INFORMATION**
	Engine components of the station wagon ... 37
PART B - COMPONENT OVERHAUL	
Renovation - general remarks ... 16	**DIAGNOSIS AND FAULT FINDING**
Crankshaft, main and big end bearings - overhaul ... 17	Scope of diagnosis ... 1
Cylinders, pistons, small ends - overhaul ... 18	Fault finding - engine will not run at all ... 2
Work on the cylinder head ... 19	Fault finding - engine runs erratically ... 3
Valve gear - overhaul ... 20	Diagnosis of knocks, noises, roughness, smoke ... 4
Oil pump - overhaul ... 21	Details of fault finding tests ... 5
Flywheel - overhaul ... 22	

Specifications

Car	New 500	Sport	500 D	500/500L	Station wagon	1973 model
Engine type number	110.000	110.004	110D.000	110F.000	120.000	126.000
Cylinders	2	2	2	2	2 (horizontal)	2
Bore (mm)	66	67.4	67.4	67.4	67.4	73.5
Stroke (mm)	70	70	70	70	70	70
Displacement (cm^3)	479	499.5	499.5	499.5	499.5	594
Bore (inch)	2.598	2.654	2.654	2.654	2.654	2.89
Stroke (inch)	2.756	2.756	2.756	2.756	2.756	2.756
Displacement ($inch^3$)	29.2	30.5	30.5	30.5	30.5	36.2
Compression ratio	7 : 1	8.6 : 1	7.1 : 1	7.1 : 1	7.1 : 1	7.5 : 1
Net power, bhp	16.5	21	17.5	18	17.5	23
at rpm	4400	4700	4400	4600	4600	4800
Gross power, bhp	21	25	22	22	21.5	26
Torque (net): lbf.ft	20	25	22.4	22.4	21.7	29
kgf.m	2.8	3.5	3.1	3.1	3.0	4.0
at rpm	3500	3500	3500	3000	3000	3400

Crankshaft

Main journal diameters	In	mm
Standard	2.1248 - 2.1256	53.97 - 53.99
0.2 mm undersize	2.1169 - 2.1177	53.77 - 53.79
0.4 mm undersize	2.1091 - 2.1098	53.57 - 53.59
0.6 mm undersize	2.1012 - 2.1020	53.37 - 53.39
0.8 mm undersize	2.0933 - 2.0941	53.17 - 53.19
1.0 mm undersize	2.0854 - 2.0862	52.97 - 52.99

Crank pin diameters	In	mm
Standard	1.7328 - 1.7336	44.013 - 44.033
.010 in undersize	1.7228 - 1.7236	43.759 - 43.779
.020 in undersize	1.7128 - 1.7136	43.505 - 43.525
.030 in undersize	1.7028 - 1.7036	43.251 - 43.271
.040 in undersize	1.6928 - 1.6936	42.997 - 43.017

Note inches are the primary units in this case

Clearances

	In	mm
Main bearing; chain end	.0012 - .0026	.030 - .065
flywheel end	.0177 - .0032	.045 - .080
wear limit	.0039	.10
Big ends; new	.0004 - .0024	.011 - .061
wear limit	.0059	.15

Cylinder bores (standard sizes)

Engine type	110.000	110F.000, 110D.000, 110.004 120.000
Class A	2.5984 - 2.5988 inch (66.000 - 66.010 mm)	2.6535 - 2.6539 inch (67.400 - 67.410 mm)
Class B	2.5988 - 2.5992 inch (66.010 - 66.020 mm)	2.6539 - 2.6543 inch (67.410 - 67.420 mm)
Class C	2.5992 - 2.5996 inch (66.020 - 66.030 mm)	2.6543 - 2.6547 inch (67.420 - 67.430 mm)

Piston clearance	110.000	110.004	110F.000, 110D.000, 120.000
At skirt bottom and right angles to gudgeon pin (standard and oversize)	.0004 - .0012 inch (0.010 - 0.030 mm)	.0016 - .0024 inch (0.040 - 0.060 mm)	.0012 - .0020 inch (0.030 - 0.050 mm)

Piston ring gap (Top three rings)	(all engines)	(fitted in bore)	.0098 - .0138 inch (0.25 mm - 0.35 mm)
Bottom (slotted) ring	No gap		

Oversizes with gradings A, B, C, to suit bores..	1st oversize	2nd oversize	3rd oversize
	0.2 mm	0.4 mm	0.6 mm

Gudgeon pin fit in connecting rod	Bushing inside diameter (pressed and reamed)	Piston pin diameter	Pin clearance in bushing
Standard	.7874 - .7876 inch (20.000 - 20.006 mm)	.7870 - .7872 inch (19.990 - 19.995 mm)	.0002 - .0006 inch (0.005 - 0.016 mm)
Oversize 0.2 mm	.7953 - .7955 inch (20.200 - 20.206 mm)	.7949 - .7951 inch (20.190 - 20.195 mm)	.0002 - .0006 inch (0.005 - 0.016 mm)

Gudgeon pin in piston	Press fit	(installed at 90°C)	(pinch of 0 - .01 mm)

Compression test ... 100 - 107 lbf/in^2 (7.0 - 7.5 kg/cm^2) with maximum variation 5% between cylinders

Cylinder height ... 3.5433 inch (90 mm) maximum tolerance .015 mm

Piston wear limits
Maximum clearances:—

Piston to cylinder:	Piston top	.0098 inch (.25 mm)
	Skirt bottom	.0059 inch (.15 mm)
Ring in groove:	Top ring	.0079 inch (.20 mm)
	Second ring	.0079 inch (.20 mm)
	Third ring	.0059 inch (.15 mm)
	Bottom ring	.0059 inch (.15 mm)
Ring gap: installed in bore:—	Top three rings	.0197 inch (.50 mm)
	Bottom ring	Touch fit at all times

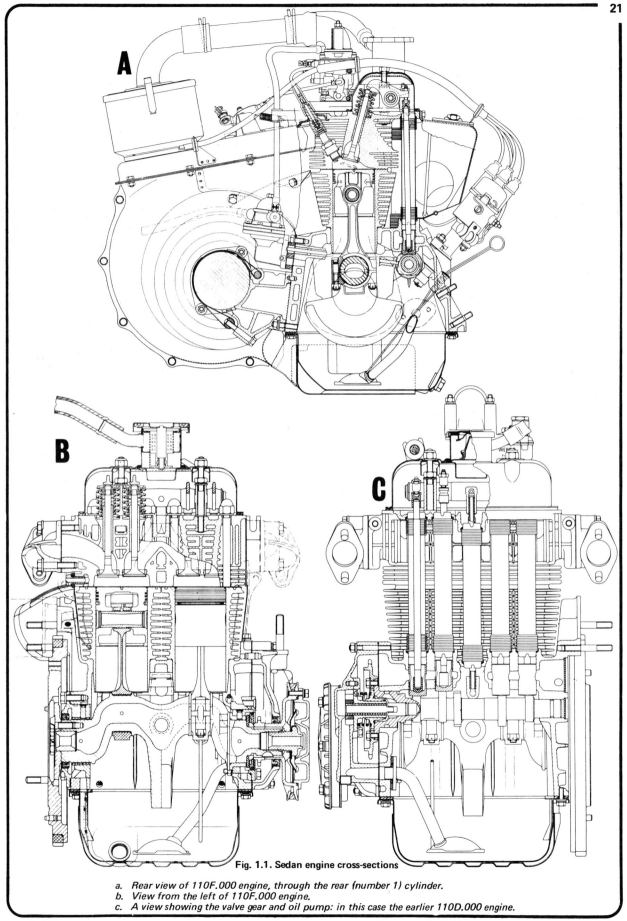

Fig. 1.1. Sedan engine cross-sections

a. Rear view of 110F.000 engine, through the rear (number 1) cylinder.
b. View from the left of 110F.000 engine.
c. A view showing the valve gear and oil pump: in this case the earlier 110D.000 engine.

Chapter 1/Engine

Valves and guides

	inch	mm
Guide hole in cylinder head	.5118 - .5125	13.000 - 13.018
Guide outer diameter	.5139 - .5143	13.052 - 13.062
Guide I.D. when press fitted	.3158 - .3165	8.022 - 8.040
Guide pinch fit in head	.00134 - .00244	0.034 - 0.062
Valve stem diameter: Inlet	.3144 - .3150	7.985 - 8.000
Exhaust	.3136 - .3142	7.965 - 7.980
Stem/guide clearance: Inlet	.00087 - .00217	0.022 - 0.055
Exhaust	.00165 - .00295	0.042 - 0.075
Maximum clearance: Wear limit	.0059	0.15
Seat angle: Cylinder head		$45° \pm 5'$
Seat angle: Valve face		$45° 30' \pm 5'$
Valve maximum diameter: Inlet	1.2598	32
Exhaust	1.0630	28
Valve seat width	.055	1.4

Valve springs

	Number of coils	Free length Inch	mm
110.000	7.75	1.8898	48
110.004	8.75	2.2519	57.2
110D.000	8.75	2.2519	57.2
110F.000) Inner	8.5	1.5827	40.2
120.000) Outer	6.5	1.8465	46.9

Camshaft bearings

	Camshaft journal diameter	Bore diameter	Fit clearance
Chain end	1.6919 - 1.6929 in (42.975 - 43.000 mm)	1.6939 - 1.6954 in (43.025 - 43.064 mm)	.0010 - .0035 in (0.025 - 0.089 mm)
Flywheel end	.8653 - .8661 in (21.979 - 22.000 mm)	.8669 - .8682 in (22.020 - 22.053 mm)	.0008 - .0029 in (0.020 - 0.074 mm)

Tappets/cam followers

	Tappet diameter	Bore diameter	Fit clearance
Standard	.8660 - .8653 in (21.996 - 21.978 mm)	.8670 - .8663 in (22.021 - 22.003 mm)	.0003 - .0017 in (0.007 - 0.043 mm)
Oversize .0020 in (0.05 mm)	.8680 - .8672 in (22.046 - 22.028 mm)	.8689 - .8682 in (22.071 - 22.053 mm)	.0003 - .0017 in (0.007 - 0.043 mm)
Oversize .0040 in (0.10 mm)	.8700 - .8692 in (22.096 - 22.078 mm)	.8709 - .8702 in (22.121 - 22.103 mm)	.0003 - .0017 in (0.007 - 0.043 mm)
Wear limit	.0032 in (.08 mm)		

Valve timing

	110.000	110D.000, 110F.000, 110.004, 120.000
Special timing clearance: Inlet	.018 in (.45 mm)	.015 in (.39 mm)
Exhaust	.015 in (.38 mm)	.015 in (.39 mm)
Inlet opens: BTDC	9°	25°
Inlet closes: ABDC	70°	51°
Exhaust opens: BBDC	50°	64°
Exhaust closes: ATDC	19°	12°

Valve clearances

Tappets set cold (early 110.000 engines)	.004 in (.10 mm)	.006 in (.15 mm)
Valve lift (110.000 engine): Inlet	.326 in (8.28 mm)	
Exhaust	.324 in (8.24 mm)	
(110.004 onwards): Both	.360 in (9.15 mm)	

Valve rockers

Clearance rocker to shaft: New	.016 - .053 mm
Wear limit	.15 mm (.0059 in)
Clearance shaft to support: New	.005 - .035 mm
Wear limit	.10 mm (.0039 in)

Tightening torques

	lbf. ft	kgf. m
Main bearing support to crankcase	15	2.1
Flywheel to crankshaft	23	3.2
Connecting rod big end cap	24	3.3
Rocker shaft to cylinder head	15	2.1
First pass	18	2.5
Cylinder head nuts	24	3.3
Camshaft chain sprocket	7	0.9
Oil filter cover on pulley	6	8
Fan to dynamo shaft	25	3.5
Pulley to flange - dynamo	15	2.0
Exhaust elbow to head	18	2.5
Sparking plugs	19 - 21	2.5 - 3.0

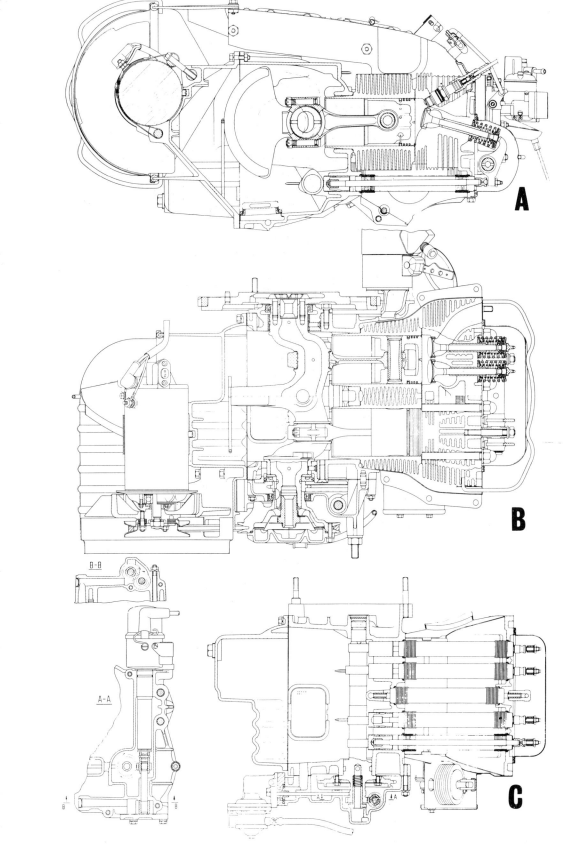

Fig. 1.2. Station wagon engine cross-sections

a. Rear view of 120.000 engine.
b. Plan view
c. 120.000 engine: Section views through valve mechanism, fuel and oil pumps, and distributor.

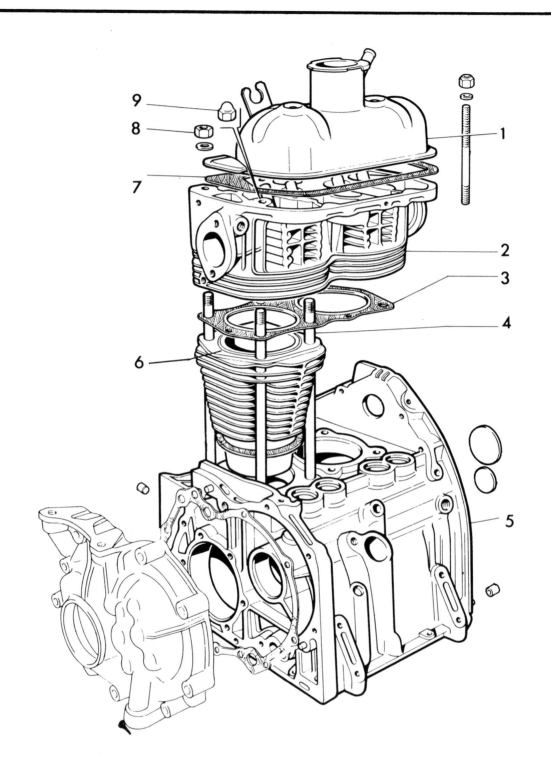

Fig. 1.3. Sedan cylinder parts

1. Rocker box
2. Cylinder head
3. Cylinder head gasket
4. Cylinder head studs
5. Crankcase
6. Cylinders
7. Rocker box gasket
8. Open cylinder head nuts (4)
9. Capped cylinder head nuts (4) under rocker box

Chapter 1/Engine

1 General description

The engine is a two cylinder air cooled four stroke. The two cylinders are in line, and in the case of the various "Sedan" versions of the car, are upright. For the station wagon a special version, the type 120.000 is made with the cylinders laid over to the right so that it can fit under the floor. The two cylinder barrels are separate individual cast iron castings, and with their fins are very similar to those of a motor cycle. They share a common aluminium cylinder head. This is held to the crankcase by eitht long studs, with the cylinders clamped between head and crankcase. The crankcase is bolted up to form one unit with the transmission casing that houses the gearbox and differential. All these casings are of aluminium to minimise weight, especially important as the power unit is at the rear. They are die cast, so have a good finish and in production can be made to tight tolerances with minimum machining.

The camshaft is in the crankcase and works the valves through push rods.

The cylinder head, like the cylinder barrels, has large fins for cooling. The cooling air is guided over these by a cowling that shrouds the whole engine. The air is blown by a centrifugal fan (axial for the station wagon) mounted on the dynamo shaft. The dynamo and thus the fan is driven by a rubber belt from a pulley at the rear end of the crankshaft.

The oil filter works by centrifugal action, and is in the pulley that drives the fan belt.

The car was first introduced with a capacity of 479 cc. At that time there was a "Sports" variant, and this included in its special specification, an engine of 499.5 cc. Later the more powerful engine used in the "Sports" was adopted for all cars. Various detail changes have been made at short intervals, but with the introduction in 1965 of the type 110F the engines became recognisably different. One of the main changes for the 'F' version was the fitting of a channel in the joint between cylinder head and barrels down which any leaking exhaust could go, lest it reach people in the car through the heating system. An improved crankcase breathing system was fitted, in conjunction with a larger air cleaner. Double valve springs were adopted. As the clutch was changed to the diaphragm spring type, the flywheel had to be modified.

At the end of 1972 the 594 cc engine of the new and larger 126 car was adopted for the 500. Apart from the enlargement of the cylinder bore this engine is the same basic design.

The type 120.000 engine of the Station wagon uses the same rotating parts as the upright engine. But because it is laid over on its side a different crankcase, sump, and air cowling are needed. The carburettor and distributor are different too.

2 Planning work on the engine

1 This Chapter covers the sequence for a complete engine overhaul. Most owners will not need to do such drastic things. To tackle some particular job the component removal will be found in the early sections, renovation in the middle, and reassembly at the end of the Chapter.
2 The majority of work can be done with the engine in place, and a lot of labour is thus avoided. Contrariwise the removal is not all that difficult or lengthy, and once a number of jobs need doing at one session, it may save time in the end to take it out.
3 Removing the engine for the first time will take something like 2 hours. Someone used to the job can do it in less than an hour, unaided. Taking it out with the transmission makes the job a little slower, and dirtier, as there are a number of components to be undone underneath. The engine must be removed for work on the main bearings and crankshaft. It is recommended no attempt is made to remove the cylinders from the crankcase until the engine has been removed from the car. There is insufficient space around the front cylinder. There is risk of breaking piston rings or dirt falling into the crankcase. Otherwise all work could be done in situ. But it is considered worthwhile taking the engine out if two or more serious things have to be done concurrently, such as removing the sump, taking off the cylinder head, withdrawing the camshaft, taking off the dynamo. It must also come out to give access to the clutch.
4 The decision to take out the engine depends on how easy its "lifting" will be. The ideal way is to pull it out backwards with a trolley jack under the sump: This is how we did it. You could probably get it out by lifting with two strong men. But getting it back is difficult because the weight must be transferred to clocks or an ordinary jack underneath, and the engine slid in straight. With the engine coming out alone without transmission, it is lighter, but the clutch and gearbox shaft must be accurately lined up on reassembly, and no stress put on the latter. Taking it out complete with transmission is heavier, but nothing is damaged so easily. It is possible to move the engine on rollers, made from bits of pipe. To push the car forward away from the engine is another method, provided the floor of the garage is smooth (and there is room) and you have enough helpers for the car to be moved whilst you steady the engine on its blocks, perched on an ordinary jack.
5 Sometimes engine removal will be dictated by need to get at the transmission. It is possible to remove the gearbox leaving the engine in place, but this involves much work under the car, and will need a good ramp or pit. It is recommended that for work on the transmission the whole power unit is taken out.
6 The major influence will be the availability of the trolley jack. With it, engine removal is very simple.
7 The removal of the engine cover (bonnet) is worthwhile for quite minor tasks, and of course essential for major ones. To remove the cover undo the nut on the right hand hinge. Open the cover and unclip the lead on the left going to the number plate lamp. Support the cover on the knees, and unclip the check strap from its fixing on the central rear engine mounting pillar. Press down the right end of the strap where it goes into its slot on this mounting, and disengage the end of the wire hook from the slot. New cars may have a little plastic plug under the strap in the slot to prevent it rattling; this must be prised out first. Now the cover can be slid along the hinge and taken off. Put it down on something soft where it will not get scratched, (photos).
8 Should you decide to leave the engine in place, it could still be of help in improving accessibility to remove the rear panel across the engine compartment. To do this the engine must be supported by a jack under the sump, with a broad plank between the two to spread the load and prevent the outer air-cooling ducts on the sump being squashed. See section 4.9.
9 Accessibility can be further improved by removing all the air cooling systems cowling. This is described in section 6, paragraph 2 onwards. But it is only worth while when a lot of work needs to be done concurrently. An example is the overhaul of the dynamo under the 18,000 miles task and removal of the cylinder head at the same time.
10 Should you decide to remove the engine a point to be considered in advance is what to do with it when you get it out. Diagnosis and decisions about the clutch and transmission that you may be taking out at the same are discussed in their Chapters. At the end of this Chapter is fault finding and diagnosis on the engine. It is best to try and find out what is wrong before work starts, rather than taking it apart to see what is wrong by looking inside. You want to discuss in advance with your garage what spares you will need, or whether it is worth getting a reconditioned engine. The price of these is quite reasonable considering you will get a complete overhaul all to the makers proper standards. In October 1972 a "bare engine" was £38. This is the engine less cylinder head, clutch, starter, carburettor or petrol pump. To get the head, an "incomplete engine" is needed, costing £52. A reground crankshaft is £6.50. By the time you have added the costs of pistons and cylinders, and then other minor parts the bill rises quickly. Unless you do the job thoroughly, things like a worn oil pump could cause a failure again quite soon. Other components like the valve gear will be noisy. It could have been their clatter that made you take the engine apart in the first place. The other thing that must influence your decision as to whether you yourself overhaul the engine is what you use as a garage. The engine must be rebuilt in

2.7a. Undo the nut on the bonnet hinge

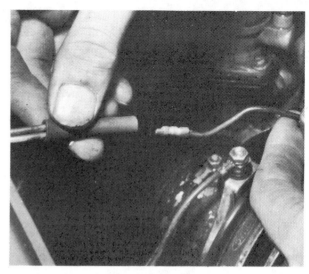

2.7b. Unplug the number plate light

2.7c. Unhook the bonnet strap

2.7d. and slide it off its hinges

Fig. 1.4. Moving parts of the engine

1 Piston
2 Piston ring set
3 Gudgeon pin
4 Circlip
5 Small end bush
6 Connecting rod
7 Big end shells
8 Main bearings in their housings
9 Gasket and seal for flywheel-end main bearing
10 Gasket and seal for flywheel-end main bearing
11 Flywheel
12 Flywheel securing bolts and tab washer
13 Flywheel securing bolts and tab washer
14 Clutch shaft spiggot bearing

a clean place free from dust. You need space to lay out the components, and where you can leave them so, undisturbed. Otherwise there is risk of them getting muddled up. Furthermore you will need to be able to borrow the accurate instruments needed to measure components to decide what to replace. You will need 2 inch and 3 inch micrometers for the crankshaft journals (or a vernier gauge) and an internal comparitor for the cylinder bores. (See sections 17 and 18).

11 The previous paragraphs discuss overhaul of a Sedan, with the upright engine. On the Station wagon the horizontal engine will need to come out in more instances. In many aspects the engine on this car is more accessible. But it gets dirtier, and is hemmed in by dirty body panels. It is also very low down, so tiring to work on.

12 One of the most frustrating delays when doing any repair is having to wait for spares. There are many items that you can be sure you will need, so can buy in advance with certainty. Others are pretty sure to be needed, and are relatively cheap, so it is well worth getting them, as the delay and travelling expense when you do find you need them could cost more. Finally there are those parts that are more expensive, and less likely to be needed. As you may have made a sizeable order of other parts, it is suggested that you ask your garage to take back any you do not use that are valuable. Check this first. Again, some parts you ought to get and fit, because with the car in bits, the joint factors of their cost and likelihood of being near the end of their useful life, make it sensible to replace them anyway. On this basis the list below has been drawn up.

13 Parts essential for reassembly without risk of subsequent problems:-

Engine gasket set, including:
Cylinder head gasket
Pushrod tube rings
Rocker box gasket
Rocker box stud washers
Carburettor upper gasket
Carburettor lower gasket
Oil filter rubber ring
Cylinder base gaskets
Valve stem seals
Sump gasket
Timing chain cover gasket
Rear main bearing gasket
Big end self lock nuts
Rubber ring: oil pick-up
Gaskets petrol pump to crankcase
Oil seal crankshaft, flywheel end
Oil seal crankshaft, chain end

14 Routine maintenance items best renewed whilst car being overhauled:

Spark plugs
Contact breaker points (pair)
Dynamo brushes (set)
Fan belt
Air cleaner element

15 Items likely to be needed, therefore worth stocking:

Exhaust valves
Tube for pushrod (replace the shortest or most battered)

16 If the engine is stripped it could well be worth fitting, even though not fully worn out:

Big end shells (pairs) (by size)
Clutch driven plate

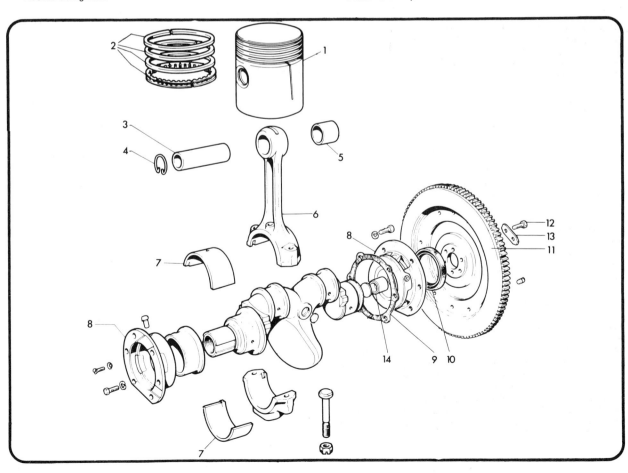

PART A - DISMANTLING

3 Removing the cylinder head in situ

1 Remove the air trunk from the air cleaner to the carburettor by undoing the clips holding on the air cleaner top and removing the two bolts holding the elbow to the carburettor. On early cars the elbow can be left in place, and instead the trunk pulled off it. Disconnect the breather from the rocker-box near the filler cap.

2 From the carburettor remove the fuel pipe, and tuck it away near the pump where dirt will not get down it. Unclamp the choke inner cable from the lever on the carburettor, and the outer from its bracket. Unclip the throttle linkage from the carburettor, at the relay lever on the air cowling. Unscrew the two nuts holding the carburettor to the engine, lift it and the drip tray underneath clear.

3 Unplug the plug leads from the spark plugs, and the low tension wire from the distributor. Remove the distributor cap. Pull the leads out of their mounting clips on the engine and stow them and the distributor cap out of the way near the coil.

4 Remove the bolt that goes upwards into the distributor to clamp it to its pedestal, and pull the distributor out of its mounting. (It cannot go back the wrong way as the drive tongue and slot on the shaft are offset).

5 Undo the two bolts on the right of the engine compartment, and take off the right apron dirt shield.

6 Take out the two studs holding each exhaust elbow to the front and rear faces of the cylinder head. Undo the four nuts holding the silencer to the side of the crankcase, and lift off the complete exhaust.

7 The air cowling must now be freed from the cylinder head. There are two bolts on the left of the cylinder head, close to the carburettor flange. Two more go into the engine from the front and from the rear. Note on the 110F engine onwards, one of those on the front and one on the rear have the drilling for the exhaust blow-by safety system. They must go back afterwards in the same holes. Then there are four bolts holding the left half of the cowling to the right. Three more bolts hold the right cowling to the cylinder head. Now pull the cowling halves away from the engine. The distributor had to be removed to give room for the right section to move.

8 Undo its two nuts and take off the rocker box.

9 From now on lay out all parts in order as you remove them so they can go back in the same place.

10 Remove the rocker shaft. This is done by releasing the two nuts on the long studs that also hold down the rocker box. The nuts must be slackened evenly to unload the valve springs. Lift the rocker shaft clear with all the rockers. Cover it with clean paper so that it will stay clean, unless the engine is old and full of sludge, in which case the rockers and shaft will have to be dismantled for cleaning.

11 Lift out the four push rods. Keep them in order so that they go back where they came from. It is handy to stick them through holes in a cardboard box so they cannot roll about and get muddled.

12 Pull out the oil pipe sticking up between the holes for push-rods to valves 2 and 3. It is a simple push fit.

13 Gradually undo the cylinder head nuts. They should be slackened in reverse to the order given in the tightening diagram, a ¼ turn at a time, until unloaded. Note that four are capped or domed nuts, and that these go on the studs under the rocker box, to prevent oil leaks.

14 Now comes the difficult bit. The head is clear for lifting, but it will be firmly stuck to the cylinders. These in their turn are less firmly stuck to the crankcase. The crankcase joint must not be disturbed, otherwise there is risk of breaking the thin gasket there. Do not turn the engine over, or the pistons will push the cylinders out of the crankcase. Hit the head from the side with a rawhide mallet to break the joint. On no account try to prise into the cylinder-to-head joint, as the surface will surely be ruined. But it is possible to lever the two apart on the cooling fins, but near their root, where they are reasonably strong. All this is difficult because the cowling cannot be pushed back very far.

15 Once the head is off the engine must only be turned over when the cylinders are held down firmly into the crankcase.

16 Lift out the five tubes surrounding the push rods and oil pipe. If their bottom seals get left behind, fish for them with a bit of wire. Do not disturb dirt around the bottom of the tube seats, lest it fall into the tappets.

17 It is recommended that on the Station wagon the engine is removed to take off the cylinder head.

18 Overhaul or decarbonising of the head is in Section 19, and its refitting in Section 32.

Fig. 1.5. Engine ancilleries
a. Sedan 110 engine mountings

1 Support under transmission
2 Rubber blocks
3 Engine mounting spring
4 Mounting bracket
5 Mounting bump stop
6 Pivot
7 Mounting swinging arm

b. Engine casings (Sedan)

1 Crankcase
2 Sump gasket
3 Sump
4 Load spreading was here
5 Sump bolt
6 Drain plug
7 Timing chain cover
8 Crankshaft oil seal

c. Mountings for the 120 engine of the Station Wagon

1 Support under transmission
2 Rubber blocks
3 Engine mounting spring
4 Engine mounting bracket

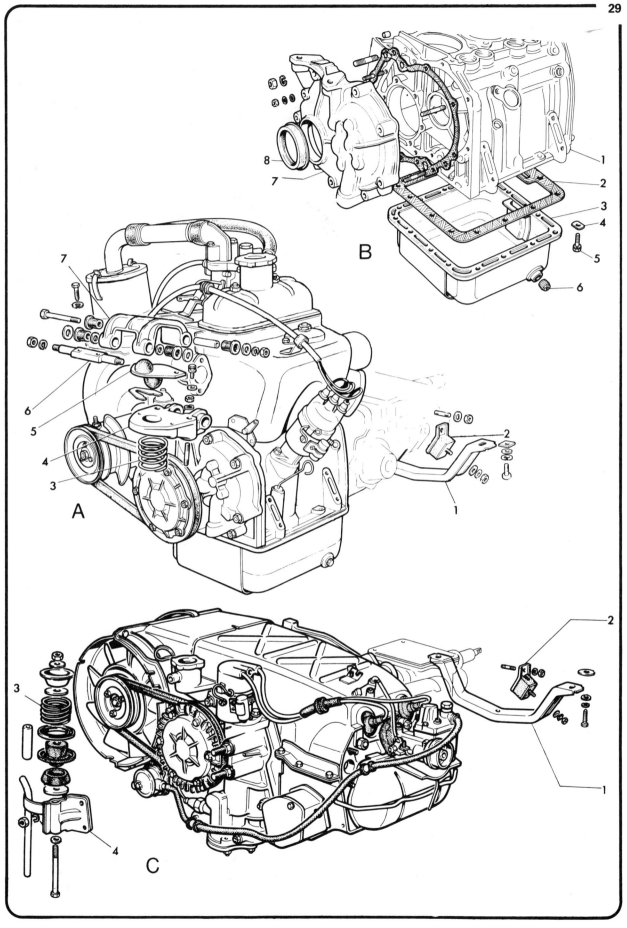

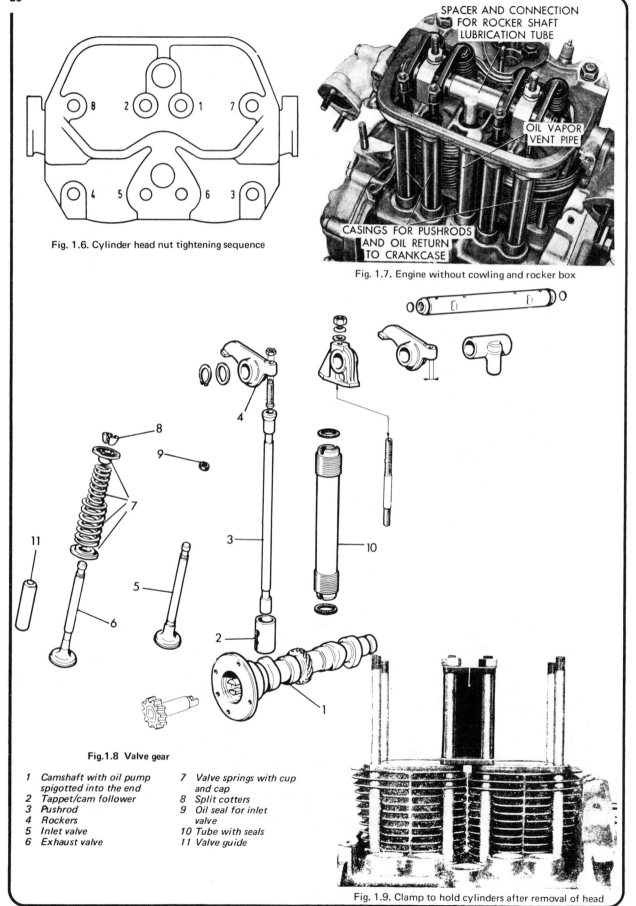

Fig. 1.6. Cylinder head nut tightening sequence

Fig. 1.7. Engine without cowling and rocker box

Fig.1.8 Valve gear

1 Camshaft with oil pump spigotted into the end
2 Tappet/cam follower
3 Pushrod
4 Rockers
5 Inlet valve
6 Exhaust valve
7 Valve springs with cup and cap
8 Split cotters
9 Oil seal for inlet valve
10 Tube with seals
11 Valve guide

Fig. 1.9. Clamp to hold cylinders after removal of head

4 Removing the engine - leaving the gearbox in place

1 At the front of the car disconnect the fuel pipe from the tank, to prevent syphoning when disconnected at the rear, (photo).
2 Uncouple the live battery lead, (photo).
3 Drain the engine oil.
4 At the rear of the car unclip the plug leads from the plugs, and remove the distributor cap, unclip the leads from the two 'U' clips on the rear of the engine, undo the king lead from the coil, the low tension lead from the terminal on the distributor and from the coil and remove all the ignition leads, (photo).
5 Remove the two leads from the top of the dynamo. The two terminal posts are different sizes, so the leads cannot be muddled up. (8 mm and 10 mm spanners), (photo).
6 Undo the clip securing the petrol pipe to the petrol pump, and slide the pipe off the union on the pump.
7 Unclamp the throttle cable from the relay lever on the air trunking (8 mm spanner). Undo the upright clamp for the outer cable from the air trunking, (8 mm) and coil the cable away to one side. Undo the choke inner cable from the lever and the outer from the clamp on the carburettor, (photo).
8 Unplug the oil pressure warning light wire from the sender on the right side of the crankcase, (photo).
9 Remove the apron covering the gap between the right hand side of the engine and the body over the exhaust (2 bolts), (photo).
10 From underneath the car remove the left hand apron. There are seven screws holding it to the side of the body and to the rear panel (photo).

11 Take the weight of the engine on a trolley jack under the sump, with a plank to spread the load and prevent damage to the double skin that forms the ducting to blow cooling air round it, (photo).
12 Remove the two bolts that screw down into the rear panel to hold the central engine mount to it. Undo the nuts on the two studs sticking up from the timing case cover, and lift the complete engine mounting bracket to take out the large coil spring (photo).

13 Remove the four nuts holding the rear panel to the side of the car body. (13 mm spanner). Note under one on the left is the earthing lead for the engine. On De Luxe cars undo the bolt either side holding the bumper extensions to the mudguard, (photo).
14 Pull the rear panel clear, (photo).
15 Put the panel where the point will not get scratched.
16 On the starter motor take out the split pin from the pin holding the cable to the operating lever. Pull out the pin. Undo the electric cable from the terminal on the starter (13 mm).

17 Undo the two nuts (three on later cars) holding the starter to the engine, and lift it out.
18 Slacken the clip and take the large cooling air intake flexible trunk off the ducting on the left of the engine.
19 Remove the cooling air outlet flexible trunk from the right of the engine.

20 Slacken the four nuts holding the engine and transmission together around the flywheel housing; the top two first. Then remove them.
21 The engine is now free, and can be drawn straight out backwards. Be sure to keep it straight so that no stress is put on the gearbox shaft in the clutch, and put a hand to it so that it will not topple off the jack.
22 On the Station wagon there are some variations. Six nuts secure the rear panel to the body. With the weight on the jack, but the jack not lifting the engine, the nuts should be removed from the studs holding the mounting to the rear of the engine. The panel can then be pulled back with the mounting still attached to it, sliding it off the studs on the engine.

5 Removing the engine and transmission complete

1 Jack up the car to a good height so that work underneath is easy. Lower the car onto firm steady blocks. Insert extra blocks in case the others are not as firm as you thought. The wheels must be clear of the ground as they need to be turned later.
2 Do all the jobs listed in the previous section for removing the engine on its own except that there is no need to remove the starter motor itself after disconnecting the electric and control cables, nor of course are the four nuts holding the engine and transmission together undone.
3 Drain the transmission oil.
4 Under the car, unscrew the speedometer cable from its connection to the transmission, (photo).
5 Remove the nut on the bolt connecting the gear linkage to the pushrod into the gearbox. Remove the special step bolt and washer. Note the way this washer fits. (10 mm nut and 13 mm bolt), (photos).
6 Remove the pull off spring from the clutch withdrawal lever. Unscrew the lock nut and the adjuster nut from the end of the cable (10 and 17 mm spanners), (photo).
7 Remove the three bolts (13 mm) from each drive shaft at both hubs, turning the wheels to get at each in turn. As the drive shaft is taken away from the hub remove the spring from inside, (photo).
8 Remove the two bolts securing the 'U' shaped cross member under the transmission to the floor beside the suspension mounting pivots. (17 mm spanner), (photo).
9 The complete power unit is now clear. Lower the jack slightly. It should be lowered enough for the front crossmember ends to pass under the rear suspension mountings. The power unit can now be drawn rearwards clear of the car. Be careful that the transmission does not catch on the control cables that are still underneath, (photos).
10 Check all the rubbers in the mountings, and note such any cracked, torn, or perished for replacement.

6 Prepatory Stripping - including air cowling and dynamo

1 Split the engine from the gearbox. Remove the nuts securing the starter motor to the top of the clutch housing. Remove the four nuts (13 mm spanner), holding the clutch housing to the engine. Then lift the transmission clear of the engine. Note that the drive shafts are still in place on the transmission. The work on this is described in Chapter 6. If not done before, drain the oil from engine, and transmission, (photos).
2 Remove the fan belt. Undo the three nuts clamping the two halves of the dynamo pulley together. Take off the spacers and half-pulley, then lift off the belt.
3 Disconnect the throttle linkage at the relay lever on the air ducting. Remove the two 13 mm bolts holding the air ducting on the rear face of the engine. Note that on the 110F engine the bolts with a hole for the safety venting for exhaust fumes. Remove the other 10 mm bolts securing the ducting round the carburettor flange and those holding the two halves of the cowling together at the front and rear of the engine. Undo the bolt at the bottom of the clamp holding the dynamo to the bracket on the side of the engine. Remove the sector of shield that is between the dynamo and the mounting block, and then pull out the pin holding the strap to the crankcase bracket. Note that there is a pip on the shield that registers with the hole in the strap. The complete assembly of the left hand air trunking and dynamo can now be lifted clear, (photos).
4 Remove the bolt securing the distributor clamping plate to the crankcase, and pull the distributor up out of the crankcase. If the crankcase is not going to be split block up the hole for the distributor with clean rag to prevent dirt going in. Remove the bolts holding the right hand air trunking to the crankcase and the cylinder head. Now lift the right hand trunking away from the engine, (photos).

4.1. Take the fuel pipe off the front to prevent syphoning

4.2. and the battery lead just in case

4.4. Remove all ignition leads including the low tension

4.5. Disconnect the dynamo

4.7. Unplug the throttle cable at the relay lever and the choke at the carburettor

4.8. Unplug the oil pressure sender

4.9. Remove the right apron over the exhaust

4.10. From underneath remove the left apron

4.11. Take the weight of the engine

4.12. Undo two bolts into the panel and two nuts holding the mounting on the engine

4.13. Undo the nuts at each side of the rear panel

4.14. and pull it off. We hadn't undone the two nuts to lift the whole mounting, so the spring made it difficult

5.4. Under the car undo the speedometer cable

5.5a. Undo the lever linkage at the gearbox. Check whilst you're there the rubber noise dampers on the rod for damage

5.5b. It is an unusual stepped bolt. Note the way it goes for reassembly

5.6. Take the spring off the clutch lever and the nut and locknut from the cable

5.7. Disconnect the drive shafts from the hubs

5.8. Undo the transmission mounting bracket under the floor

5.9a. Lower the jack slightly

5.9b. Pull the power unit clear of the car

6.1a. Take off the starter. (Three studs on later cars). When the engine is in the car this can be done reaching round the right of the engine

6.1b. Unbolt the transmission from the engine

6.3a. To take off the air cooling cowling first disconnect the throttle

6.3b. Undo the cowling from the head at the side

6.3c. and at back and front

6.3d. and the bolts holding the two halves together

6.3e. Undo the bolt under the dynamo

6.3f. Take out the shield

6.3g. Pull out the pin at the top of the dynamo strap

6.3h. The left cowling with dynamo and fan will now come off

5 Now that the engine has been cleared of the cowling the whole unit can be cleaned prior to further work.

6 Before stripping the engine, measure the value lift. See Section 20.9.

7 Removing the cylinder head (engine on bench)

1 Remove the exhaust system complete. Take out first the two bolts holding each elbow to the front and to the rear of the cylinder head. Undo the four nuts holding the silencer bracket to the side of the crankcase. Lift away the two exhaust pipes with the elbows still on the ends, the silencer and the bracket as a unit.

2 Remove the two bolts holding the rocker box and lift it off.

3 Remove the carburettor. First take off the pipe from the petrol pump to the float chamber. Then undo the two nuts securing the carburettor to the engine. Lift off the carburettor, and the base plate with drip tray, (photos).

4 Remove the rockers by undoing the two nuts on the studs that hold the rocker cover. Slacken off the nuts gradually and let the valve springs be unloaded pushing the rocker shaft up level as the two nuts are undone. Lift the rocker shaft with all the rockers clear. Lift out the four pushrods. Keep them in the same order as they were before. It is suggested that you keep them in an old cardboard box, punching four holes through the lid to keep them in their correct order. Lift out the oil feed pipe that is between rockers 2 and 3. This pipe is a simple push fit, (photos).

5 Undo the cylinder head holding down nuts. Slacken them in reverse order to that given in the diagram. Initially only give one a ¼ of a turn at a time, until all are unloaded. Finally remove all the nuts. Note that the four capped nuts come off the four studs inside the rocker box. These are needed to stop oil running down the threads, (photos).

6.4a. The distributor is in the way of the right cowling ....

6.4b. with that off and all bolts out the right section comes off

Chapter 1/Engine

6 The head is now ready to lift off. If the cylinder barrels are not going to be removed care must be taken or they will shift in the crankcase, and break their joint. The engine must not be turned over or the pistons will push the cylinders up out of the crankcase, (photo).

7 A blow from the side with a mallet should break the head joint. On no account try to prise at the joint between head and cylinder or the surface will be ruined. If jerks and thumps will not free it it is possible with care to lever near the roots of the air cooling fins, where they are strong.

8 Once the joint is broken lift the head a small way; till the studs are just level with the head surface. Check the five tubes are staying behind in the crankcase. If they go up with the head they may fall off. They are fragile.

9 Take off the cylinder head, and put it down where the bottom face cannot get damaged. Then take out the tubes, (photo).

10 Take off the head gasket, (photo).

11 If the cylinders are not being removed organise clamps to hold the barrels down into the crankcase, using odd lengths of wood and two of the cylinder head nuts.

12 Assuming you will be removing the cylinders, mark each barrel with its cylinder number and front, so they will go back in the same position and the same way round.

13 Lift off the cylinder barrels, (photo).

8 Stripping the centrifugal oil filter

1 If the engine is still in the car, and the whole filter has to be removed, that is, the pulley taken off the crankshaft, take off the rear panel as described in section 4, paragraph 9 onwards.

2 Undo the six bolts (10 mm) holding the cover on the pulley. If the engine is in the car put it in gear. If removed hold it at the flywheel end. Put some rag to catch the oil in the filter that will pour out, (photo).

3 Note the end cover will only go back in one position, the bolts being unevenly spaced, so the TDC mark will be in the correct place. Note also the rubber sealing ring, which should be renewed. Scrape out all the dirt, (photos).

4 If stripping further, bend back the tab washer holding the hollow central bolt. Remove the bolt (32 mm spanner), (photo).

5 Take off the oil thrower ring, noting that its concave side is towards the engine, (photo).

6 Pull off the pulley, (photo).

9 Timing chain removal

1 Having removed the oil filter/pulley from the crankshaft as described above, remove the chain cover.

2 There are two large nuts under the engine mounting and a ring of smaller ones all the way round. Slacken them all off evenly, not only to undo the cover gradually as would be normal, but also because of the oil pressure relief valve, (photo).

3 The oil pump is on the end of the camshaft. Outboard of it is the relief valve, with its spring held by the cover. This spring will push the cover along the studs, (photo).

4 There are little levers on the links of the chain. These are the tensioners. On other Fiats they are often on the engine side of the chain, but on the 500 they are outside. Also notice the timing marks on the sprockets.

5 Bend back the tab washers locking the bolts on the camshaft sprocket, (photo).

6 Undo the sprocket bolts.

7 Take the sprocket off the end of the camshaft, (photo) and then unloop the chain from the sprocket on the crankshaft. The holes on the camshaft sprocket are unevenly spaced, so it cannot be put back the wrong way.

8 The timing chain cover has one of the two crankshaft oil seals. This should not be reused. Drive the old one carefully out of the cover.

10 Removing the flywheel

1 Remove the ring of bolts holding the clutch cover to the flywheel, slacken them all gradually evenly and diagonally.

2 Lift off the clutch, and the driven plate.

3 Bend back the tab washers on the bolts holding the flywheel to the end of the crankshaft. Undo these bolts. To stop the engine turning over whilst these bolts are undone, stick a screwdriver into the starter ring on the flywheel, and hold it against one of the studs for the clutch cover, on the flywheel housing. (Photo).

4 Lift off the flywheel, (photo).

11 Crankcase minor components

1 Undo the two nuts holding the petrol pump to the crankcase, and lift away the pump. Note that there is a plastic distance piece between the pump and the crankcase. Pull out the long rod that works the petrol pump from the camshaft on the other side of the engine. If the bush for it on the crankcase is free, take it out to prevent loss, (photos).

2 Remove the oil pressure warning light sender unit from the other side of the crankcase.

3 Turn the engine upside down. Slacken off all the sump bolts initially a small amount. Note that under each bolt is an oblong, load spreading, washer, (photo).

4 Take off the sump. In doing so take care that the joint between the steel pressing and the alloy crankcase is not spoilt. If it has to be prised off, this must be done very carefully not to bend the steel, or deform the aluminium. The best implement is a flat paint scraper.

5 Undo the two nuts holding the oil suction pipe to the timing chain end of the engine, and remove it. Under the flange for the pipe is a spacer and between this and the crankcase a rubber sealing ring, (photo).

12 Camshaft - removal

1 It is usually impractical to remove the camshaft with the engine in the car. The cam followers must be held up to clear the cams and distributor drive gear. The rockers and push rods having been removed, their own weight would still drive them down. The cowling, cylinder head and rocker tubes being already removed, the cam followers can be pulled up by a magnet above their normal running position, where they will probably wedge. They might even come right out, upwards. However, with the engine out, it can be turned upside down to get the cam followers out of the way.

2 The distributor, petrol pump, and timing chain having been removed as described in preceding sections, the camshaft is free.

3 Pull the camshaft out of the engine.

4 Push the cam followers out of their holes in the crankcase, putting them down in order, so that each can go back in the same hole. They should come out either way. Reassembly will be easier if they will pass upwards, but they may go more easily out through the crankcase, as a lip of dirt at the top may stop them.

13 Connecting rods and pistons

1 Since the sump has now been removed the big ends are accessible. Slacken evenly and then undo the self locking nuts clamping the big end bearing caps. Throw away the nuts; new ones must be used. Lift off the caps. Note the cylinder marking number on the side nearest the camshaft, (photo).

2 Holding the assembly by the piston take the connecting rod out of the crankcase. One the side of the big end on the connecting rod is the cylinder number again, matching that on the bearing cap, (photo).

3 From the cap and the rod take the two halves of both big end

7.3a. Take off the rocker box. Undo the carburettor

7.3b. Lift it off,

7.3c. and the drip tray

7.4a. Undo evenly the two nuts holding down the rocker shaft and take it off

7.4b. Take out the rockers keeping them in order,

7.4c. and the rocker oil pipe

7.5a. Slacken the head nuts gradually in reverse order to the tightening sequence in the diagram

7.5b. The four capped nuts go under the rocker box

7.6. Lift off the head making sure the cylinders and tubes stay behind

7.9. Take out the tubes

7.10. Lift off the gasket

7.13. Don't disturb the cylinders unless you have to. Mark them if you do to show number and direction

8.2. Undoing the oil filter

8.3a. As the cover comes off the sealing ring is revealed. A bit of oil will drip out

8.3b. The car we bought had been neglected. Do not let so much dirt build up as it may break away and get into the bearing

8.4. Bend back the tab washer and undo the nut

8.5. After the hollow bolt and tab washer comes the concave oil thrower: Note the way round it goes

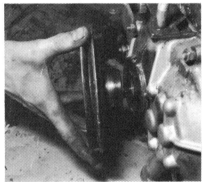
8.6. Take off the pulley

9.2 Slacken all the nuts evenly to ease the cover off

9.3. The cover compresses the spring of the oil pressure relief valve

9.5. Bend back the tab washers and remove the bolts on the camshaft sprocket

9.7. Take off the sprocket and unloop the chain from the crankshaft. The timing cannot be lost

10.3. After removing the clutch undo the tabs and take out the bolts

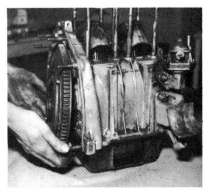
10.4. Lift off the flywheel

11.1a. Undo the petrol pump

11.1b. Take it off with the plastic spacer

11.1c. and pull out the push rod

11.3. Take off the sump

11.5. Between the oil pick-up pipe and the crankcase is a spacer and sealing ring

13.1. The big end numbers are on the cam shaft side of cap and rod

bearing shells. Do not prise them, but slide them round in the 'U' of the bearing. Mark each one with ordinary pencil on the back so that it will go back in the same position.

4 If the pistons are not being replaced do not take them off the connecting rods. However to remove them take out one of the circlips in the piston. Push the gudgeon pin out with the fingers. If it does not come easily immerse the piston for about half a minute in a saucepan of boiling water. Aluminium expands more than steel, so this frees the pin. Put the pin down in such a way that you will remember which way round it goes.

5 Note that the piston was fitted on the connecting rod with its slot on the same side as the number on the big end.

14 Crankshaft and main bearing - removal

1 Pull the timing chain sprocket from the end of the crankshaft. This must be kept square or it will jam. Some come easily, but a stiff one will need a proper hub puller to pull it square. Any levering must be done evenly; both sides together, (photo).
2 Prise out the woodruff key locating the sprocket on the shaft, and put it in a safe place like a jam jar.
3 If they come easily at this stage take out the distance piece with spring oil retaining ring, and the second distance piece, behind the sprocket. Their removal now can help main bearing removal; contrariwise their own extraction is much easier when the bearing is off.
4 The main bearings are one piece rings held in housings in the end faces of the crankcase.
5 Slacken the screws holding them, evenly and gradually. Note the plates are different, and that on the one at the timing chain end there are two countersunk screws. These two screws go underneath the run of the chain; under their heads are conical lock washers.
6 The bearings are difficult to remove from the crankcase.

They can be prised out using two levers at opposite sides, but be careful not to damage the soft aluminium crankcase. They must be moved very evenly, or the bearing will jam. (photo).
7 The timing chain end is complicated by the oil retaining ring and the distance pieces. The whole assembly has to come a long way along the crankshaft, and must be kept square, or it will jam, (photo).
8 The one at the flywheel end can be removed by placing the engine on end, timing chain end down, so that the end of the crankshaft takes the weight. This then pushes the bearing housing at the other end out. Be careful; but the studs should take the weight once the bearing moves before the shaft goes so far the crank throws damage the bearing seating, (photo).
9 Once both bearings are out, push the crankshaft along inside the crankcase until the balancing web is up one end, then lift the other end out of the hole in the crankcase, and clear, threading the shaft out.
10 Do not let the crankshaft tumble about in the crankcase in case the seats for the main bearing houses are burred, (photo).

15 Stripping the oil pump

1 The oil pump is mounted on the inside of the timing chain cover. It is driven by a slotted drive on the end of the camshaft. The pressure relief valve is worked by a disc and spring concentric with the drive shaft. To remove the spring and disc, compress the disc against the spring and take off the circlip on the end of the oil pump drive shaft, (photos).
2 The bolts holding the oil pump to the timing chain case are now accessible, and can be undone. Slacken them all evenly a little at a time, and then remove them.
3 Lift the oil pump cover off the spindle. This will then show the driving spindle driving one gear, and the driven gear wheel. There are no gaskets, (photos).

13.2. Take the pistons and rods out after removing the cylinders. Reassembly is different

14.1. A proper pulley may help getting this off. Don't loose the woodruff key

14.6. Those screwdrivers are pressing on the cover face and may damage it

14.7. In the housing of the chain-end main is a distance piece with oil retaining ring, and behind that another plain distance piece

14.8. With the other main out the shaft will hang from the flywheel end. The shaft may drive the bearing out without having to use levers

14.10. If the crankshaft falls about in the crankcase it may put burrs in the bearing seats

15.1a. The oil pump in the chain cover driven by the camshaft

15.1b. Compress the spring to get off the circlip

15.1c. Then take off the pressure relief valve

15.3a. The oil pump

15.3b. is simply

15.3c. dismantled

4   When reassembling the pump lubricate all parts thoroughly with engine oil.
5   The pump can be removed with the pressure relief valve still in place, using a spanner rather than a socket to get at the bolts behind. However, the pump cannot then be stripped properly for cleaning. The pump cover has a dowel to locate it in the chain case.
6   In reassembling the pressure relief valve note that there is a tongue on the valve which must be aligned with a cut out in the cover of the oil pump, otherwise the spring cannot be compressed to put on the circlip.

PART B - COMPONENT OVERHAUL

16 Renovation - general remarks

1   With the components stripped they can be thoroughly cleaned and then examined. If the car has been run on cheap oil it will be covered internally in sludge. All this must be washed off, and out of hollow sections and oil ways. If the engine is clean inside having been run on high quality detergent oil do not immerse components in dirty baths of solvents, less dirt is washed in. Scrape off all remnants of gaskets from all joint surfaces. Use a blunt paint scraper.
2   A decision must be taken on what must be replaced. This could be due to cracks, scoring, or just wear. If things you can measure or see are bad, then this will be indicative that other components less easily assessed are in the same state. However if things are not too bad the engine could be given an extension of life by replacing the components subject to the highest wear; the pistons with rings and the main and big end bearing shells. The valves will certainly need regrinding, possibly both exhaust valves need renewing, and the cylinder head refacing, even all three! However such a partial refit will not last long if such things as the crankshaft and cylinders are badly worn, as they will be worn oval, and the new components will suffer quickly.
3   Some measurements need micrometers, vernier gauges and the like. On others where clearance is the vital factor, feeler gauges can achieve a lot. If the feeler set is taken apart individual blades can be inserted in things on their own, to see how great a thickness can be put in before the components become stiff to move.

17 Crankshaft, mains, big ends - overhaul

1   The bearing surfaces of the crankshaft journals and pins should be bright and smooth. If there are scratches or scoring they will need regrinding. Measure the diameters of the bearings in a number of directions, looking for ovality. If the ovality exceeds 0.001 in - 6.03 mm) then this is excessive purely as ovality, but also implies that overall wear will be too much. Take it to a FIAT agent, who can arrange the regrinding simultaneously with the supply of the main and big end shells to the suitable undersize. Otherwise you must take it to a machine shop, who could advise you, and then you order the new shells.
2   If the crankshaft ovality seems alright you may be able to measure the clearance to confirm overall wear is within the condemnation limit. It is difficult to measure. However, crankshaft wear is usually indicated by the ovality: No ovality means negligible wear. The shells can be assumed to have worn. The wear limits are given in the specifications at the beginning of the Chapter.
3   Look on the old ones for their serial numbers to confirm their size when reordering. If you are not having the crankshaft reground this will confirm whether you have a standard or undersize crankshaft.
4   The main bearings are supplied already pressed into their housings.
5   Because the bearings are so easily fitted and are relatively cheap it is false economy to try and make do with the old ones.
6   If the white metal of the old shells is badly broken up, and if the engine has been knocking badly, and for a long time, then this will be confirmation that the crankshaft needs regrinding as well.
7   The connecting rods need to be straight. It is difficult to check them without proper instruments and blocks on a surface plate. If the engine is just suffering "fair wear and tear" then straightness can be assumed. But they should be checked if there has been any catastrophy such as seizure. If you have not the experience or equipment, take the connecting rods to your local engineering works. They can be bent straight if faulty. Should a connecting rod be replaced check that the rods are within 2 oz (6 grams) weight of each other. If necessary shave metal off the heavier rod, keeping a smooth shape, from the wide portion where the shoulders run down to the bosses for the bolts.
8   Whilst the crankshaft is out the spigot bearing for the quill shaft through the clutch to the transmission is accessible. Now is the time to remove the old one should it need replacement. See Chapter 6.

18 Cylinder, piston, small end - overhaul

1   The oil consumption and exhaust oil smoke will have given some indication as to the wear of the bores and pistons.
2   Clean the cylinders thoroughly inside and out. Caked or hard lacquered dirt left on the cooling fins spoils the air cooling a lot. At the cylinder bottom is a taper to guide the piston rings into the bore. This wants to be clean and smooth so that they will slide in easily; emery paper may be needed. Scrape the carbon off the unworn lip at the top of the bore so that its original size can be compared with the worn.
3   Measure the bore diameters. They will be worn more near the top than the bottom, and more across than fore and aft. If the difference between the largest and smallest dimension exceeds .006 in (.15 mm) then the ovality is excessive and a rebore is necessary. If the bores have any scores they should be rebored; if one has to be done, the other must too to match it.
4   Check the cylinder height from its seat on the crankcase. If too long the compression ratio will be too low; this fault is unlikely, but some previous owner may have machined off some metal to raise the compression ratio, or the barrels may already have been near the limits and crept slightly. If too short there is risk of carbon building up on the piston and causing a foul. If they are too short, they must be replaced; again both, not just one.
5   Even if the cylinders may not need reboring it is likely the pistons and rings will need replacing. They will have worn on their outer circumferential surfaces, and where the ring contacts the piston land in its groove.
6   Slide an appropriate feeler sideways into the piston groove to measure the clearance between each ring and its neighbouring land (and write down the result to think about later).
7   Carefully expand the rings and lift them off the piston. Insert the piston into the cylinder at its correct axis (slot towards camshaft side, ie right). Find the fattest feeler gauge that will pass between piston and bore with the piston halfway down its stroke to get the widest part of the bore. Also measure opposite the ridge at the top. You have now got the actual clearance, at the worn bit and the cylinder wear. Take out the pistons. Insert a piston ring. Push it halfway down the cylinder with a piston, so that it is square. Measure the gap in the ring.
8   If the clearance between cylinder and piston is excessive then the pistons must be replaced, and new rings fitted to them. New rings can be fitted to old pistons by specialist firms who will machine out the grooves, which will be worn conical, and supply suitable fat rings. But this is not really economic. The wear limits are listed separately in the specifications.
9   Note that it is most important if fitting new pistons in the existing bores that the top ring has a step cut out of its top so that it will not hit the ridge left at the top of the bore. This will have been left by a worn piston and ring. Should normal new rings be fitted, the foul would anyway cause a knock, but probably also break the rings. Note also that the second and

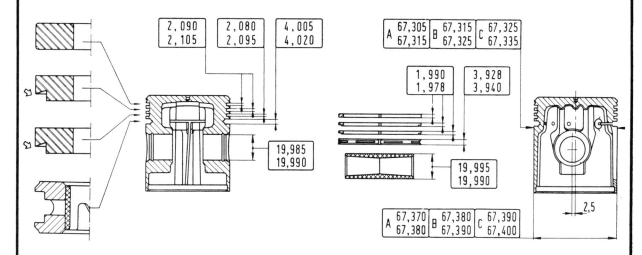

Fig. 1.10. Dimensions of pistons, rings, and gudgeon pins, in mm, for 110D000 and 120.000 engines

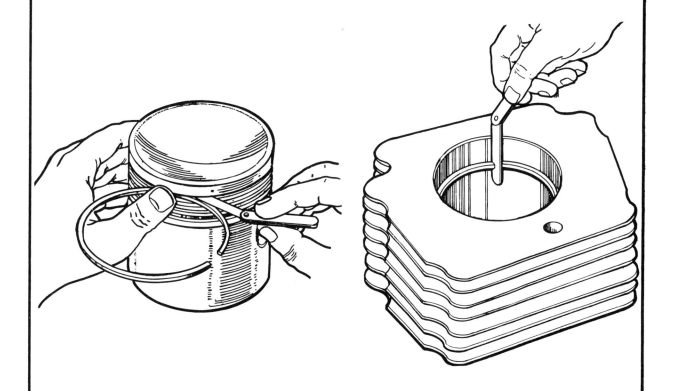

Fig.1.11 Measuring ring clearance in piston groove

Fig.1.12. Measuring piston ring gap in cylinder bore

third rings have a special scraping bottom edge. The original pores have three sizes, A, B and C, which is stamped on the cylinder top.

10 FIAT supply new pistons (and oversize ones) complete with rings and gudgeon pin; but not ones with the stepped top ring necessary if not reboring. This may persuade you to have the cylinders rebored. It would make a better job anyway.

11 If the pistons need replacing then the gudgeon pin may be worn too. Also there is likely to be wear in the small end bush in the connecting rod. If the pistons are being replaced it would be nice to do the same for these bushes and fit new gudgeon pins of standard size. However it is difficult to fit the bushes, as they must be pressed in, then reamed to size, and finally an oilway machined in the top. If you do have the facilities the instructions are as follows:

Press in the new bush into the connecting rod. Then ream it using an adjustable reamer so as not to oversize, to 22.000 mm. Then cut an oilway in the bush to match the slot in the connecting rod. This should be milled using a cutter of diameter 55 mm and width 3 mm. Cut until the centre of the milling cutter is 35 mm away from the centre of the small end bush. Oversize gudgeon pins are available; but then the old bush must be reamed out, again using an adjustable reamer to take it out to the new size. The new pistons will come with the standard size small ends, so their reaming presents another problem. Note that their bearing is smaller than that in the connecting rod bush to allow for their expansion. It is suggested that the new bushes and standard gudgeon pin is the best choice. Your FIAT agent should be able to fit and ream them for you.

If you have to cut the oilway yourself without a miller you might be able to drill it if you are careful. Before starting look at the old oilway. Then using a 3 mm or 7/64 in drill make two holes at the extremities of the slot, pointing "inwards" so that the hole is funnel shaped. Break through very gently to prevent tearing out bearing material.

It would help to have a dowel of hardwood in the bush on which to press down. Then drill as many more holes as you can fit; perhaps two more, between the original two. Chisel out the remaining parts between all the holes, using a sharp screwdriver in default of a proper chisel. If you have a fine file clean up the hole with that, but probably you will have to make do with a sharp penknife and emery paper torn into narrow ribbons.

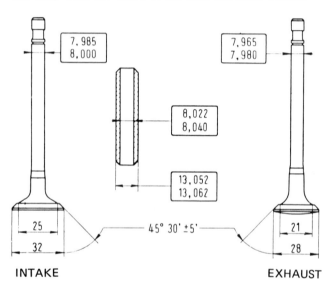

Fig. 1.13. Dimensions of inlet and exhaust valves and guides

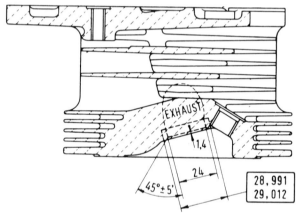

Fig. 1.14. Dimensions of exhaust valve seats

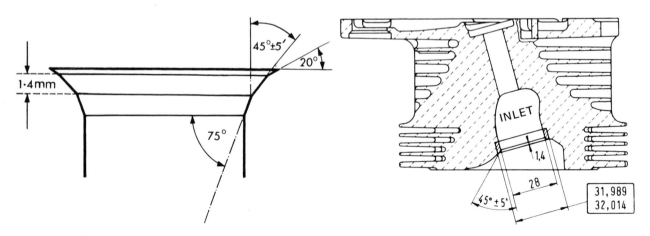

Fig. 1.15. Valve seat shape

Fig. 1.16. Dimensions of inlet valve seats

19 Work on the cylinder head

1 The overhaul of the cylinder head is much the same whether it has been taken off with the engine in place for a "top overhaul", or if part of a more general engine rebuild. But if doing only a "top overhaul" then it is likely that the soundness of rocker gear and valve guides can be taken for granted.

2 Clean off the general dirt and oil from the cylinder head.

3 Prepare a box for the valves and their retainers. The lid of a cardboard box can have four holes punched in it to hold the valves. All must be labelled so they go back where they came from (see Fig.1.8).

4 Place a valve spring clamp round the head and compress the spring enough to fish out the two parts of the split cotter with a small screwdriver. Release the clamp. Take off the cap, the outer (and inner on later cars) valve spring, and the rubber sealing ring. (Early cars have a snap ring as well as the split cotters, but sealing rings only on the inlet valves).

5 Examine the valve seats for signs of bad pitting, and in the case of the exhausts, burning. Check the mating surface of the head to the cylinder for signs of gasket blowing. With the carbon still on the head the washing marks of leaks and blows should be apparent. Such a little engine is usually worked hard. The carbon should be dry and fairly thin, and look "hot"; maybe white, but anyway greyish. If thick, damp, and soft it indicates too much oil getting into the cylinders, either up past the pistons, or down the inlet valve guides.

6 If the cylinder head gasket has been blowing (see also Chapter 2), then the head will need refacing. Either your FIAT agent or a motor cycle repairer will get this done for you. The minimum amount necessary to get a clean flat surface should be removed.

7 If the valve seats are badly pitted or burned they will need refacing. Again your FIAT garage, or any large repairer will have the cutters. If you try to do it by lengthy valve grinding then the valve will get badly worn, and the seating contact area will be too wide. The refacing operation includes narrowing of the seat with cutters at 20° and 75°.

8 If the valve seats are being faced then the valves could be refaced too by the same firm at the same time. However, if the head is alright but the exhaust valves bad, then the most convenient and economical thing to do is to buy two exhaust valves. The inlet valves are usually in quite good condition.

9 The valve guides will be worn. It is very difficult to measure the wear. A useful yardstick is that if you need the crankshaft regrinding you will need new valve guides. It is tricky pressing the old ones out and the new in. It is recommended you get the FIAT agent to do it. He also will have the experience on which to judge the wear. The guides wear more than the valve stems so fitting new valves will not help this much.

10 Having decided what work must be done by a professional, now clean up the head. Scrape off all the carbon. Be careful not to scratch the valve seats. These are hard inserts, but a small scratch will be difficult to grind out. The head is made of aluminium, so soft, and easily cut when scraping. The combustion chambers, inlet and exhaust ports must be cleaned. A blunt screwdriver and flat paint scraper are useful. If using a wire brush on an electric drill, wear goggles, (photo).

11 It is after this that the head should be taken for any machining. Also during the cleaning any cracks will be found. Should this unlikely event occur the solution must be another head.

12 Clean all carbon off the valves. It is convenient to do their head tops by putting them (unfixed) in their seat in the cylinder head. Scrape off all deposits under the head, and down the valve stem. The rubbing surface where the stem runs in the guide should be highly polished by wear; do not touch this, but the part of the stem nearer the head may have lacquered deposits that can be removed with fine emery paper. At this stage do not touch the valve's seating surface. Clean out the ducts for the gasket leak safety by-pass system.

13 Now grind in the valves. Even new ones will need grinding in to bed them to their actual seat. If the seats and valves or just the one, have been recut, the hand grinding must still be done, (photo).

14 The idea is to rub the valve to and fro to mate valve and seat, and give a smooth flat perfectly circular sealing surface. The end product should be matt grey, without any rings or shine worn on it. The seating surface should be about midway up the valve's 45° surface, not at the top which happens if a valve is refaced so often it becomes small, and sits too deep in the seat.

15 The best tool is a rubber sucker on the end of a stick. Unless the sucker is good, and the valve absolutely oil free it keeps coming off. Handles that clamp to the stem overcome this, but they are clumsy to hold. On no account use an electric drill; a to-and-fro motion is essential.

16 If the valves and seats have been refaced you will only need fine grinding paste. If cleaning up worn seats start with coarse.

17 Smear a little of the paste all round the seat, being very careful to get none on the valve stem. Insert the valve in its place. Put the valve grinding handle on the valve, and pushing it lightly down onto its seat, rotate one way then the other. Every now and then lift the valve clear of the seat, turn it about half a turn, and then carry on. By altering the position the grinding paste is redistributed, and also the valve will work all round the seat and make it circular.

18 If coarse paste is used try and judge the change to fine just before all marks have disappeared so that they and the large grain of the coarse paste are ground out at the same time; the least metal rubbed off the better, otherwise the seat will get too broad.

19 The seat should be a uniform pale grey. Rings are a sign that the valve has not been lifted and turned enough. If a long grind is needed the paste will get blunt, so wipe off the old and smear on some new. A spring under the valve head can help in the lifting, but it is difficult to find a suitable light one.

20 Clean off all traces of valve grinding paste very thoroughly. Wipe out the valve guides by pushing clean rag through a number of times. Engine oil makes a good detergent for this, particularly if squirting through hard with a good oil can such as a Wesco. Leave everything oily to prevent rust (photo).

21 The valve springs may need replacing. Measure their height as they stand free. If they have shortened by 1/16 in or 1.5 mm they should be renewed.

22 Reassemble the valves to the head. Use new rubber stem sealing rings (and put the valve into the correct seat; in which you ground it!). Oil the valve guides, and the valves all over, before assembly.

23 Insert the first valve in its seat. Put the spring cup over the stem followed by the springs, and locate them round the guide. If the springs have a varying spiral put the end which the spring coils closest together next to the head. Put the cap on the spring. (Photos).

24 Put the valve spring compressing clamp round the head and compress the spring; it needs to go just so far that the groove in the end of the stem is about half clear of the cap. Push the sealing ring over the end of the valve stem, and push it down to the bottom of the narrow part. Put in the two split cotters. Undo the clamp gradually, if necessary moving the spring cap about to let is slide up the cotters to clamp them properly, (photo).

25 If you removed the cylinder head just for a top overhaul the cleaning of the head must be matched by removing the carbon from the piston crowns. Turn the engine over, whilst holding the cylinders down, to get the pistons to top dead centre. Scrape, using a flat burnt paint scraper or wide screwdriver, all the carbon and odd bits of gasket off the piston crown and cylinder top face. Clean out the groove which is the gasket blow-by safety passage. When all is clean debris that has fallen down between the pistons and cylinders must be removed. Turn the engine over to lower the piston about an inch. Carefully wipe away carbon sticking to the walls, rubbing gently so as not to knock off the carbon on the top, unworn bits of the wall. This reputedly should be left as it helps the piston seal at TDC. Now squirt engine oil over the piston to flood the edge. Work the

19.10. Scrape the carbon off and clean up the head and valves

19.13. Keep lifting and moving the valve to another sector, so it is ground free of rings on the seat

19.20. Clean the valves, especially the stems and guides, very thoroughly; then oil them

19.23a. Put on the cap

19.23b. Then springs and cap

19.24. Compress the spring till the stem just sticks out, and put on the rubber seal, and push it down with the cotters

piston up and down several times (holding the cylinders down every time) to wash the debris out, and finally leave the cylinder walls well lubricated.

20 Overhaul of the valve gear

1 The valve gear from timing chain to rockers works hard, particularly on such a small engine, which can expect to spend much time at high speed. If worn the many components can create a lot of noise which spoils a car when driven gently, and can be rather upsetting at speed. The wear would have to be extreme before failure occurs. More likely loss of power might intrude first if the camshaft is badly worn. Then also once bad wear, would become more rapid; components would be rattling about in their bearings, they would loose oil faster than the reduced flow in the valve gear could replace it, and lower the overall pressure. The wear limits are quoted in the Specifications.
2 Rockers: If clean due to use of a good oil and timely oil changes, the rockers should be left assembled when doing a top overhaul. At general overhaul they should be examined. Remove a circlip at one end. Take each component off, putting them so that they will go back in the same place. Then put each rocker and pedestal back on singly and check for wear. Be careful to get the component in its proper position. If the bushes and shaft are worn to the condemnation limit then replace the lot. Rebushing is not recommended, as the tip of the rocker will be worn too. This makes it difficult to accurately measure valve clearances. Once through the hardened outer layer of metal, wear is more rapid, so regrinding the tip is not recommended. The end of the adjuster that sits in the pushrod should be smooth and shiny. Pump oil through the oilways of the rocker shaft with a strong oil can, blocking all but one of the exits. Oil the bushes, and reassemble the oil pipe union, pedestals and rockers onto the shaft and refit the circlip.

3 Push rods: Check that the rods are straight. See that the working surfaces at both ends are shiny and smooth. Faulty ones should be replaced.
4 Cam followers/tappets: Check the bottom surface of the tappet is bright and shiny where it is pushed up by the cam. The stress is high, and gives rise to pitting, particularly if a poor quality oil has been used. Replace any that have any pits. A disintegrated tappet is very bad for the camshaft. Check the fit of the tappet in the crankcase. If the slop is up to the wear limit the oversize ones should be fitted. FIAT agents have reamers to open out the holes. If they are unobtainable an adjustable one will have to be used. If allowed to run beyond the wear limit the rate of wear will increase, as the sideways load on the tappet will tip it over. Burrs or dirt should be cleaned off the top of the holes in the crankcase ready for reassembly, so that the tappets can be fitted quite late in the reassembly process.
5 Sprockets: When new the teeth are symetrical, cut to an involute profile. Most of the wear will be on one side. Once worn so that the teeth look assymetrical replacement is due, The teeth will be through their case hardened layer so they wear rapidly. The hooked teeth wear the chain too.
6 Timing chain: If the sprockets need replacement then the chain must be changed too. If the rollers are visibly ridged it should be condemned. Otherwise the amount of sideways slop and endways movement needs experience to judge. If the valve gear has been noisy, and there has been a sort of rushing noise, then change it.
7 Camshaft: The cam profile must be checked by measuring the valve lift. This is given in the specifications as the axial movement of the valve with the tappet clearance set normal, .004 in (.10 mm). If the lift is .035 in (.9 mm) less than new the camshaft should be replaced. Otherwise power will be lost, and again this is a component which has its best wearing metal on the outside. If the camshaft bearings in the crankcase are badly worn reclamation is not economic as FIAT neither sell the undersize

bearings nor have the boring tools for opening out the crankcase to fit them. The snag to running with very worn camshaft bearings will be the loss of oil pressure in the feed to the valve gear. As FIAT do not consider any reconditioning necessary there is no option but for you to do the same. Luckily, and no doubt this is why the bearings are like this, wear seems very little.

The distributor drive teeth should be examined. Again experience, as for the timing chain, is the guide in judging whether the camshaft needs replacing for this. The cams should be smooth and shiny. If there are nicks these can be polished out with a fine abrasive, but it is important to know how the mark was made or more damage may result. But if there are deep cuts, as opposed to shallower ones that can be polished, the camshaft should be replaced.

8 Push rod tubes: The five tubes for the push rods and rocker oil feed should be checked for cracks or dinges, particularly in the concertina sections at the end. It is probable that at least one will need replacement. If the tubes are shortened they will not seal properly at the ends, and give oil leaks. Use new seals at the two ends.

21 Oil pump overhaul

1 The oil pump should only be suspect for wear at very high mileages, or after some abuse, which would be indicated by such things as the crankshaft needing regrind to the second undersize.
2 The specifications and wear limits for the pump are difficult to come within inch limits due to small tolerances expressed originally in metric measure.
3 If wear is bad the two gear wheels need replacing. In extreme cases the pump cover plate and the timing chain cover will need replacing.
4 The backlash between the gears must not exceed .20 mm.
5 New gear wheel width is 10.00 to 9.978 mm. If it is less than 9.95 mm they must be replaced.
6 Side clearance between gears and cover should not exceed .12 mm. This is not a common problem, but after high mileages may occur. In addition to the gears the chain cover with single feelers packed inside to see what size will fit yet the pump still turn.
7 The clearance between the boss on the pump cover plate for the gear shaft and the inside of the pressure relief valve that seats on it should not exceed .15 mm. If necessary the cover plate and valve must be replaced.

22 Flywheel overhaul

1 There are two things to check; the clutch pressure surface, and the starter ring.
2 If the clutch has been badly worn, or badly overheated by slipping the surface on which the clutch presses may be scored or cracked. This would wear a new clutch plate rapidly. The flywheel should not be skimmed to remove these, but replaced. This should be a fairly safe purchase from a car breaker.
3 Wear on the starter gear ring should not be bad, as the starter is the pre-engaged type. Check that there are no broken teeth, or burrs. If there is a bad defect a new flywheel is required. Minor blemishes can be filed off.

PART C - REASSEMBLY

23 Preparation for reassembly

1 This is the stage when cleanliness is vital. The work area must be clean, with no risk of dust blowing about. The surface should be covered in a few layers of clean newspaper.
2 The components should be collected in order of assembly. All old ones replaced should be put away so there is no risk of muddle.
3 All parts must be well lubricated, so that when the engine first starts the relatively rough and tight new parts will not be harmed, remembering that it will be some seconds before the oil pump fills the empty oilways and supplies pressure. An oilcan filled with engine oil will be necessary throughout assembly.
4 All tools should be laid out in order ready for quick use. They must be clean, or grit will be transferred into the engine. The tools should include a torque wrench.
5 Clean, fluff-free, rags will be needed for wiping components and your hands, that may have attracted some dirt.
6 In general, all components must be tightened gradually evenly and diagonally, to pull them squarely into place. All nuts or bolts should have some locking system, either spring washer, self lock washer, locking tab, lock nut, or a self-lock nut. All locking washers or tabs must be in good condition. Self-lock nuts are unsafe if used a second time.

24 Crankshaft and main bearing assembly

1 Fit the new oil seal to the flywheel end housing. It should be fitted with its lips facing inwards. It has to be driven in by hammering, but not with direct blows. If the old seal is relatively undamaged it makes an excellent intermediary. Otherwise use a soft faced hammer. Fit the paper gasket to this housing, smearing it lightly with grease to hold it in place.
2 Oil both main bearings and the shaft journals.
3 Thread the crankshaft into the crankcase, not letting it hit anything. An assistant will be useful to hold it.
4 Put on the chain end main bearing, turning it so that the flat part of the rim is lined up with the bottom face of the crankcase. Tap its housing into its seat in the crankcase, just as far as it easily goes at this stage; enough to hold the crankshaft.
5 Now very carefully thread the other bearing assembly over the flywheel end of the shaft. The oil seal is vulnerable. If the crankshaft is kept up the chain end of the case the seal will not have to climb onto the shaft until the bearing housing is seated in the crankcase, and all held square. Again line up the flat on the bearing housing with the crankcase bottom face, (photo).
6 Tap both bearings housings fully home, carefully end evenly. Check nothing impedes their movement, but before they are fully home, as soon as they will reach, insert all the bolts and screws into their threads. Once the housings are home, tighten the bolts, using the torque wrench (these bolts are not done up at all tight). Remember at the chain end the conical washers under the countersunk screws, in the cutaway parts, (photo).
7 Move the crankshaft to check it turns freely.

25 Connecting rods, pistons, cylinders, big ends - assembly

1 It is assumed all clearances, gaps and fits of old and new parts have been checked during the overhaul stage.
2 Fit the rings to the pistons. Slide them carefully over the crown with the least possible stretch, and steadying your fingers on the piston to keep them close and prevent jerks. Make sure they are in the right order (so put the bottom one on first), and the right way up, remembering the oil scraping cuts underneath the second and third, and if you are fitting a stepped top one, the step on top.
3 Fit No. 1 piston to No. 1 connecting rod, and if old parts, using the original gudgeon pin. The slit in the piston has to be the same side as the cylinder number on the big end. Put one of the circlips in the piston small end. Then heat the piston in a saucepan of boiling water for about two minutes. Oil the gudgeon pin and small end bush in the rod. Take the piston out of the water, shake drops off it, hold it for a moment to dry, squirt oil into its bushes then quickly line it up with the connecting rod and push the gudgeon pin and along until it reaches the circlip at the far end. It will quickly cool and the pin become too stiff to move. Now fit the second circlip, (photos).
4 Turn the piston rings round in their slots so that their gaps are on the far side of the piston from its slit, (photo).
5 The piston slit is on its lightly loaded side; the thrust is the

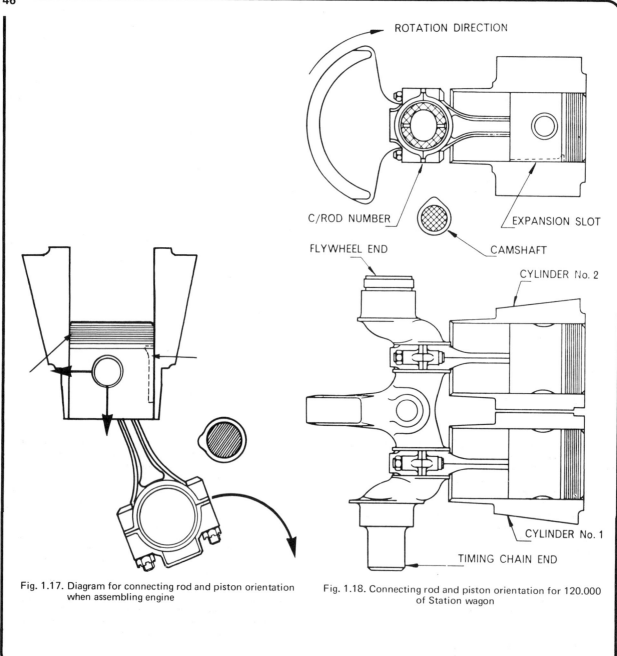

Fig. 1.17. Diagram for connecting rod and piston orientation when assembling engine

Fig. 1.18. Connecting rod and piston orientation for 120.000 of Station wagon

24.5. Line up the flats of the main bearing housings with the sump face

24.6. The bolts go into the aluminium case, so only tighten them to the specified torque, and use spring washers

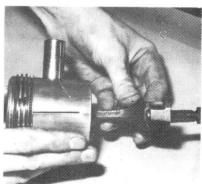

25.3a. Piston slit and number on big end go the same, camshaft, side.

other. The gaps in the piston rings will be deep into their grooves as the piston is pushed over against the cylinder wall.

6 The next thing is to put the piston into the cylinder. Doing a complete overhaul these are off the crankcase; this is easier. Oil the cylinder bore. See from the markings which is the "camshaft side", then put the cylinder on the bench upside down. Oil the piston and rings. Keeping the rings correctly placed, now invert the piston and enter its crown into the bore at the correct orientation of slit on the camshaft side. Hold the piston firmly because of the lopsided weight of the connecting rod above, and get it square to the cylinder. Lower it into the cylinder till the first ring enters the taper. Squeeze the ring with the fingers; push onwards gently pulling back the fingers as the skin gets trapped by the ring. Slide the piston on into the bore, and repeat for the other rings. If the rings do not enter the bore readily; stop. You will need some sort of clamp. It depends on your fingers, but the rings are fragile, and at this stage it would be a great nuisance and delay getting another new ring on its own, (photos).

7 If only a partial overhaul is being done then the rods will be still on their big ends in the crankcase. This means the cylinder must be lowered onto the piston. This is more difficult. The piston tends to wobble about on the connecting rod, so it is very difficult to hold it square. In this case a piston ring clamp is even more valuable. Use a proper one, or else a 3 inch hose clip of the "Jubilee" type. Tighten this gently; just enough to make the rings small enough to fit into the bore; yet loose enough to let the piston push out of the clip, as the cylinder is slid on, (photo).

8 Place a new gasket round the base of the cylinder, lightly coating it in grease, (photo).

9 Prepare the big ends for reassembly. Wipe the shells and their seats in the rod and bearing cap. All must be clean and free of fluff or oil Slide the shells into position. It will be seen that the shells stick out a little proud at the bearing faces. This is so they will be nipped firmly into place when the cap is tightened.

10 Oil the crankpins and the bearing surfaces of the shells. Oil was not wanted behind the shell to allow them to seat in snugly.

11 Fit the assembly of cylinder, piston and connecting rod to the crankcase, again checking the piston slit and big end numbers are to the camshaft side. Seat the cylinder barrel down on the crankcase from above. Then from below put the big end on the crank pin, (photos).

12 Fit the big end cap and new self locking nuts. Tighten evenly to the specified torque, (photos).

13 Hold that cylinder down onto the crankcase, and turn over the engine as a check.

14 Repeat for the second cylinder.

26 Crankcase minor components - reassembly

1 Refit the oil suction pipe with its distance piece and a new rubber ring.
2 Smear grease on the new sump gasket and fit it to the crankcase.
3 Put on the sump. Under the bolts should be the load spreading and the spring washers. Tighten evenly and diagonally.
4 The engine is now easier to work on as it can be stood on the sump.
5 Refit the oil pressure sender unit to the side.
6 Fit the petrol pipe long actuating rod (and its bush if this was a loose fit and came out).
7 Fit the first gasket, then the distance piece, second gasket, and then petrol pump. Grease both gaskets.

27 Flywheel - refitting

1 If a new spigot bearing is being put in the end of the crankshaft for the quill shaft through the clutch install this (see Chapter 6).
2 Put the flywheel on the end of the crankshaft.
3 Fit the bolts with the tab washers underneath.
4 Tighten the bolts to the specified torque, and bend over the tabs to lock them.

28 Oil pump - reassembly

1 Oil the two pump gear wheels liberally and put them in their recess in the timing chain cover.
2 Put on the oil pump cover (there is no gasket). A dowel locates the cover.
3 Fit and tighten evenly the bolts.
4 Fit the pressure relief valve, aligning its tongue with the groove in the pump cover boss, (photo).
5 Put on the PRV spring and disc. Compress the spring by hand, and fit the circlip.

29 Camshaft

Oil liberally and put the camshaft into its bearings on the crankcase, (photo).

30 Timing chain - reassembly

1 Lubricating each as they are fitted, put the first distance piece on the nose of the crankshaft, and then the second with its oil sealing ring: the plain side should be outermost (photo).
2 Fit the woodruff key in its slot, and then slide on the chain sprocket.
3 Without the chain, put the camshaft sprocket temporarily in place, and bolt it up loosely, (photo).
4 Turn the crankshaft and camshafts to get their sprocket marks lined up.
5 The marks should be close to each other, and in line with the shafts' centres.
6 Remove the camshaft sprocket again. Now oil it and the chain, and fit them, looping the chain over the crankshaft sprocket. The little tensioning levers on the sprocket should be on the outside.
7 Fit the bolts with their lock tabs to the camshaft, tighten them and bend over the tabs, (photos).
8 Fit the new oil seal to the chain cover. Tap it in gently and evenly with a soft hammer, the lips of the seal being inwards. If you have not got a soft hammer use some intermediary such as the old seal.
9 Fit the chain cover gasket over the studs on the end of the crankcase.
10 Fit the cover. Tighten all the nuts evenly to draw the cover down level as the PRV spring is compressed.

31 Oil filter - reassembly

1 Fit the fan belt pulley to the crankshaft.
2 Put on the oil thrower ring, with its concave side towards the engine.
3 Put on the tab washer, then screw in the hollow bolt. Tighten the bolt to the specified torque. Bend over the tab washer to lock it.
4 Fit a new rubber ring and then put on the oil filter cover to the pulley, making sure all is quite clean.
5 Tighten the bolts evenly and diagonally.

32 Cylinder head - replacement

1 Have a final wipe clean of the head, the piston crowns, the tops of the cylinders, and the cylinder bores. Put a little clean oil on the bores.
2 Lubricate and insert the cam followers/tappets into their holes in the crankcase.
3 Fit the new cylinder head gasket, smearing it lightly with grease, (photo).
4 Fit new seals to both ends of the tubes for push rods and oil pipe, and install the four pushrod tubes. Put the rocker oil pipe

25.3b. Heat the piston in boiling water and push the gudgeon pin before it cools

25.3c. Now fit the circlip

25.4. Put the ring gaps on the piston thrust side (no slit) spread over about 20°. Oil them

25.6a. To be sure of not breaking rings, for which you have no spares, use a clamp (or a large hose clip)

25.6b. Oil and invert the cylinder. Push the piston from the clamp into the bore

25.7. If the crankshaft and big ends have not been disturbed put the cylinders onto the pistons: hold the latter to stop them wobbling

25.8. Stick a new gasket to the cylinder with grease

25.11a. Lower the cylinders into place, in the right position; as you originally marked them; big end number and piston slit nearest the camshaft

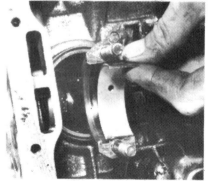

25.11b. The shells must be clean and dry when fitted in the conrod

25.12a. and the cap

25.12b. But liberally oil the bearing as the conrod and cap are fitted round the crankpin

25.12c. Tighten the nuts evenly, gradually, and finally to 24 lb ft (3.3 kg m)

28.4. Tighten the flywheel nuts and lock their tab washers

29.1. Oil the camshaft generously

30.1. On the crankshaft goes the plain distance piece, then the one with the sealing ring, plain side out, followed by the woodruff key and the sprocket

30.3. Before putting on the chain put the sprocket on alone and turn the shafts into alignment

30.7a. The bolt holes are unevenly spaced so so it only goes one way

30.7b. It should end up with the marks on the sprockets in line and the chain tensioning levers outward. Now bend over the lock tabs

32.3. Fit the gasket

32.4a. New seals to the tubes

32.4b. The push rod tubes to the crankcase

32.4c. Put the oil pipe in place followed by its tube

32.5a. When the head is about to reach the tubes pause to guide them in

32.5b. Don't force them head down or the tubes will be shortened. Once the head is down all should turn stiffly indicating all are the same length

on its hole in the block, and then add its tube, (photos).
5 Lift on the head. Lower it down the studs till these just about to appear out of the holes in the top surface of the head. The springiness of the studs will probably hold the head there. Check all the tubes are straight, and lined up with their holes in the head. Lower the head slowly making sure the tubes don't foul, as they are delicate, and could be shortened by heavy pressure. The head should go down easily, so do not force it. Once the head is sitting on the gasket try to turn the tubes round in their seat. All should be equally stiff. If any are more free than the others it indicates they are shorter. Lift off the head again, and stretch the short tube. This can be done by working the end of the tube with a thumb; push the end of the tube to one side; turn the tube around and push again, gently easing the concertina section out bit by bit all the way round. Repeat for the other end of the tube. Then try assembling it again, (photos).
6 Fit the cylinder head nuts, with their flat washers underneath. The four capped nuts go under the rocker box. Tighten the nuts gradually, in the sequence shown in the diagram. With the torque wrench bring the tightness first to 18 lb ft. Then go round again for a second pass to bring them to the final specification tightness of 24 lb ft., (photo).
7 Turn the engine over to see that it can rotate properly.
8 Block up the carburettor hole with clean rag.
9 Fit new spark plugs.
10 Insert the pushrods in their original positions.
11 Guide their bottom into the seat in the tappet.
12 On the rockers undo the locknuts and slacken the adjustment by screwing the adjusters as far as they will go into the rockers. These will make it easier to bolt down the shaft.
13 Put the rockers in place on the head, guiding the top of the oil pipe into place in the centre, and sitting all the rocker screws in the pushrods. Fit the flat and the spring washers, then tighten down the nuts to pull the shaft down parallel to the head as the valve springs are compressed.
14 Adjust the tappets. See the 6,000 mile task in Routine Maintenance, item 9, (photo).
15 Fit the new rocker box gasket. As elsewhere, do not use jointing compound, but grease may be useful to hold it in place. Check carefully that it is under the rim of the cover all the way round. Once it has been compressed, it will go back more readily after subsequent removals. Fit new sealing washers and then the nuts to the two studs.
16 Refit the exhaust. See Chapter 2. Use new gaskets between the elbows and the head. First tighten the nuts and bolts finger tight. Then tighten those nuts holding the silencer to the crankcase properly, and last do the bolts for the elbows to the head.
17 Refit the carburettor, using new gaskets between the head and the drip tray, and that and the carburettor. If the aircleaner elbow has been removed from the carburettor, cover the air intake with clean rag, (photo).
18 If the head has been removed with the engine in place in the car complete reassembly. Refit the cowling on both sides of the engine, not forgetting the bolts hidden behind. Check the two drilled bolts for the exhaust blow by safety system are in the right places. Connect the fuel pipe and throttle linkage. Refit the choke inner and outer cables. Check that the lever in the car moves that on the carburettor through its whole range.
19 Refit the distributor. It is assumed the points were renewed and set previously. Now reset the timing. See Chapter 4. Refit the distributor cap and leads. Fit the air cleaner. Refer to the starting procedures in Section 36.

33 Air cowling and dynamo - refitting

1 Fit the right section of the cowling, installing all bolts finger tight.
2 Fit the air duct to the sump.
3 Fit the dynamo to the left section of cowling, tightening those bolts. Install the left section of cowling on the engine. Again leave all bolts finger tight.
5 Check the drilled bolts for the exhaust blow-by safety system are in the correct holes.
6 Fit the sector of shield between the dynamo and its mounting, locating the pip in its hole. Fit the dynamo strap, pin at one end, and underneath the bolt, with the earthing strip.
7 Now tighten all the cowling bolts.
8 Fit the fan (see Chapter 2).
9 Fit a new fan belt. Initially put four spacers between the two halves of the pulley, and the one ring outside, and fit the spring washers and nuts. Check belt tension and adjust as necessary.
10 Connect the petrol pipe from the pump to the carburettor, and the throttle linkage from the relay lever on the left cowling.
11 Fit the distributor. It is assumed this was overhauled separately and the points have already been set. Now set the timing (see Chapter 4).

34 Replacing the engine - less transmission

1 Refit the clutch to the flywheel (see Chapter 5). It is imperative the clutch plate is accurately centred.
2 Put the gearbox in neutral.
3 Put the wooden blocks on the trolley jack and the engine on the blocks.
4 Move the engine into place. Adjust the height of the jack and tilt the engine to get it accurately lined up.
5 Slide the engine in. Turn the crankshaft over using a 10 mm spanner on the oil filter bolts on the pulley to allow the quill shaft splines to enter the clutch, whilst pushing the engine back. Push the engine right onto the transmission. Take care no weight goes on the quill shaft, moving the engine about gradually if it does not appear to be properly lined up.
6 Fit the nuts holding the engine to the transmission with flat washers and spring ones. Tighten evenly.
7 Install the rear panel. Fit the nuts to the studs, remembering the earthing strap from the engine on the left.
8 Refit the spring to the panel, and then put the rear mounting assembly on the studs on the engine. Fit and tighten those nuts. Then bolt down the swinging arm to the rear panel.
9 Lower the jack and pull it clear.
10 Refit the starter motor to the engine.
11 Now follow the next section, Section 35 from paragraph 11 onwards.

35 Replacing the engine - with transmission

1 Refit the clutch to the flywheel (see Chapter 5). Ensure the clutch driven plate is correctly centred and the right way round. Tighten the bolts gradually and diagonally.
2 Offer up the transmission to the engine. Slide the quill shaft through the clutch, turning the transmission slightly to allow the splines to enter. Fit the flat and the spring washers, and then the nuts, tightening evenly. Fit new mounting rubbers to the mountings ready for installation.
3 Fit the starter motor.
4 Put the wooden blocks on the trolley jack, and lift the complete power unit onto them. Raise the jack and slide the unit into position. Adjust till it is at the right height. Check no controls are tangled up with the front end of the transmission underneath.
5 Underneath fit the two bolts holding the bracket under the transmission to the floor, leaving them finger tight for now.
6 Install the rear panel. Fit the nuts to its studs remembering the earthing strip from the engine on the left, (photo).
7 Refit the spring to the panel, and then the rear mounting assembly to the studs on the engine. Fit and tighten the nuts on it. Let the weight almost completely off the jack. Bolt down the swinging arm of the mounting to the rear panel, (photos).
8 Go under the car again. Tighten the bracket to the floor.
9 Still under the car reconnect the speedometer cable, ensuring it is clean. Grease the threads and reconnect the gear linkage and clutch cable. Refit the clutch pull-off spring. Adjust the clutch pedal free play (¾ inch). Put molybdenam disulphide grease

32.6. The capped nuts go under the rocker box. All the nuts must be tightened in the sequence, and gradually

32.14. Fit the rockers and adjust the tappets before you forget. .006 in (.15 mm)

32.17. New gaskets under and above the drip tray

35.6. Putting back the engine is the reverse of taking it out. Don't forget the earthing wire to the panel stud on the left

35.7a. Put in the coil spring, tighten the mounting to the engine: Then take most of the weight off the jack and let the spring hold it.

35.7b. Then fit the bolts holding the mounting to the panel

35.9 When coupling up the clutch cable and spring set the free play at the pedal to ¾ inch

35.15. Remember the bumper extensions on the 500L

36.1. All should now be back. Is it? Check carefully

36.5. Start up and check for leaks on top and underneath

36.8a. Put on the bonnet

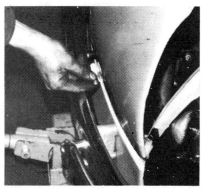

36.8b. and go for a short, gentle, test and recheck

inside on the splines of the shaft in the coupling, insert the little spring, and reconnect the drive shafts, (photo).
10 Lower the jack and pull it clear from under the engine.
11 Reconnect the starter electric cable.
12 Connect the starter control wire. Put the pin in the appropriate hole in the lever to remove lost motion, and fit the split pin.
13 Connect the air trunks to the outlet on the right and the inlet on the left.
14 Refit the right apron (two bolts).
15 Under the car refit the left apron. On De Luxe cars bolt the bumper extensions to the mudguards (photo).
16 Plug the oil pressure warning light wire into the sender.
17 Refit the throttle cable to the relay lever. Pull the inner cable out to ensure the pedal is at the closed position. Get an assistant to press the pedal when you have fixed it whilst you check at the carburettor that full throttle is obtainable.
18 Reconnect the choke cable. Adjust so that the lever in the car gives the full range of movement at the carburettor.
19 Refit the petrol pipe into the pump.
20 Connect the two leads to the dynamo.
21 Fit all the ignition leads and the spacers on the plugs. Wipe the inside of the distributor clean. Fit the rotor arm and the distributor cap.
22 Fill the sump with oil. Check the transmission was filled with oil before fitting.
23 At the front reconnect the battery and the petrol pipe out of the tank.
24 Jack up the car, and take out the supporting blocks, and lower onto the wheels.

36 Final assembly and starting up

1 All should now be ready for the moment of starting up. Check:
    No bits left over!
    Oil level and fuel
    Carburettor initial settings (Chapter 3)
    Battery in good order
    Fan belt and tension
    A general look at all nuts, bolts, wires etc, (photo).
2 The engine will have to turn over quite a time on the starter to pump fuel up to fill the float chamber. As soon as it starts check it is running smoothly and keep it a bit above idle speed; but not fast. Check oil and charging warning lights have gone out.
3 With it still running on the "choke", go quickly to the rear and check there are no disastrous leaks, particularly from the oil filter or fuel system.
4 Set the throttle stop to give a fast idle, and then close the "choke".
5 Listen to the engine for exhaust leaks; or nasty noises. Watch for lesser leaks, (photo).
6 Run for ten minutes then adjust idle for correct settings. Switch off.
7 Check all nuts and bolts for tightness, especially:
    Cylinder head nuts, in usual order and correct torque
    Exhaust
    Carburettor
    Cowling
    Engine mountings to panel and the cross member under the car
    Panel mounting to body
    Battery terminals
8 Refit the bonnet and connect up the rear number plate light, (photos).
9 Go for a short road test. Then again recheck for leaks and tightness, on the same components.
10 Allow the engine to cool and check the tappets.
11 The engine will now need running in. This involves light load and the avoidance of high speed whilst components wear off their roughness and "shrug" into place. As they are a bit tight, they might get too hot at speed. Check for leaks and looseness. Change the oil at 400 miles to wash out odd bits of dirt and abbraded metal from the running in. Gradually increase over the first 800 miles the work you ask the engine to do.

PART D - ADDITIONAL INFORMATION

37 Special instructions for station wagons

1 In Part A of this Chapter the special points to watch when removing the station wagon engine were mentioned, but the work on the engine described in Parts B and C assumes the more common upright engine. The major components of the horizontal one are all the same, so the overhaul procedures for them are similar, save for the slightly less convenient shape of the crankcase.
2 The auxiliary drives to the oil pump and distributor are different. The distributor sits on top of the timing chain cover, on a vertical shaft driven by a gearwheel on the end of the camshaft. The oil pump is at the bottom of this vertical shaft. The pressure relief valve is in the end of the camshaft, so the same instructions for removing the chain cover apply.
3 The petrol pump is on the left end of the chain cover. It must be removed before the cover, so that its actuating rod will be disengaged from the camshaft. As in the upright engine there is the long pushrod across from the pump to the camshaft. When replacing the cover take care the teeth of the vertical shaft engage in the camshaft gear wheel.
4 The rocker box is held on by a single large spring wire clip.
5 The exhaust pipes lead straight from the roof of the cylinder head, without the intermediate elbows.
6 Preparatory stripping is different due to the different cowling for the air cooling system, the fan and the generator. See the diagrams in Chapters 2 and 8.

DIAGNOSIS AND FAULT-FINDING

1 Scope of diagnosis

1 Though nominally part of the engine Chapter, diagnosis and fault-finding cannot be dissassociated from the problems of components the subject of other Chapters. The matter is therefore covered most fully here, and only narrowly in the other Chapters.
2 The word 'diagnosis' is used to refer to the consideration of symptoms of major mechanical problems, such as noises implying expensive repair or overhaul is needed.
3 'Fault-finding' implies the tracing of a defect preventing some component from functioning.
4 Defects can often be cured by luck. At other times there is no defect, merely a foolish mistake has been made. For example the engine may now start because the rotor arm has been left out, but proper diagnosis or fault finding requires knowledge as to how the thing works, and its construction. Experience helps a lot, for then symptoms can be recognised better. Symptoms must be considered and tests made in an orderly and logical way, step by step, to eliminate possibilities. Beware the dogmatic reputed expert. The true expert is usually non-committal until proved correct by actually finding the faulty component or effecting a cure. You need patience.
5 Many obtuse defects defy diagnosis by garages as they cure themselves temporarily when the car gets there. The owner who cures his own has a great advantage over the garage mechanic as he lives with the symptoms. He knows how everything has been functioning in the past; and may have some item on his conscience, such as plugs overdue for cleaning.
6 The subject is dealt with as follows:
    Fault-finding - engine will not run
    Fault-finding - engine runs erratically
    Diagnosis of knocks and noises.

2 Fault-finding - engine will not run at all

1 Problems in this section will occur under two main circumstances: either when you come to start up the engine initially; or when previously running satisfactorily.

2 Under these circumstances there are many possibilities, so the elimination system in the diagram should be followed.

3 Stoppages on the road have been found from large samples of breakdowns to be most often an ignition defect. The diagram therefore at an early stage aims to eliminate the fuel system.

4 Failure to start from cold is usually a combination of damp with dirt, weak spark because of overdue maintenance of the ignition system, and a weak battery.

5 Therefore in deciding to treat the car's temperament as a 'defect' may be misleading. On a cold damp day it is often best to try a push start before going into the fault-finding sequence. The slightest lack of verve in the way the starter spins the engine should therefore be interpreted in the chart as 'starter cranks sluggishly'.

6 The fault-finding chart is adjacent. In it reference is made to various tests. These are listed after the diagnosis tables.

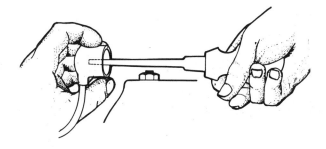

Fig. 1.20. Using a screwdriver to check for a spark at the distributor cap, if the king lead cannot be detached

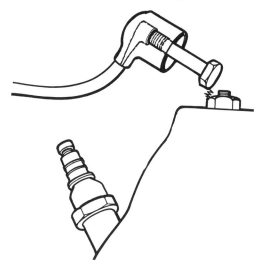

Fig. 1.19. Checking ignition HT by testing for a spark from a plug lead fitted with a shroud, using a ¼ in bolt as a probe

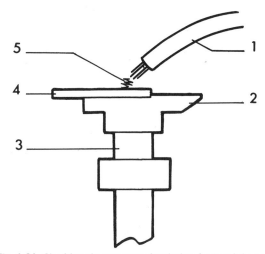

Fig. 1.21. Checking the rotor arm insulation for breakdown.

1 King lead
2 Rotor arm
3 Distributor spindle
4 Metal contact
5 There should only be one small spark as the metal is charged

Fig. 1.22. If the oil filler cap valve jams shut, pressure will build up in the crankcase. It can blow down the dipstick. If the valve jams open there is excessive drawing off of oil vapour through the carburettor. Replace the filler complete with valve when worn

Fig. 1.23. A stethoscope is useful for jobs such as differentiating between a leaking exhaust flange, and a blown cylinder head gasket venting through the drilled safety bolt

Engine will

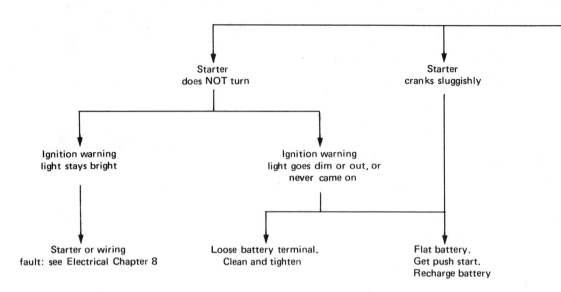
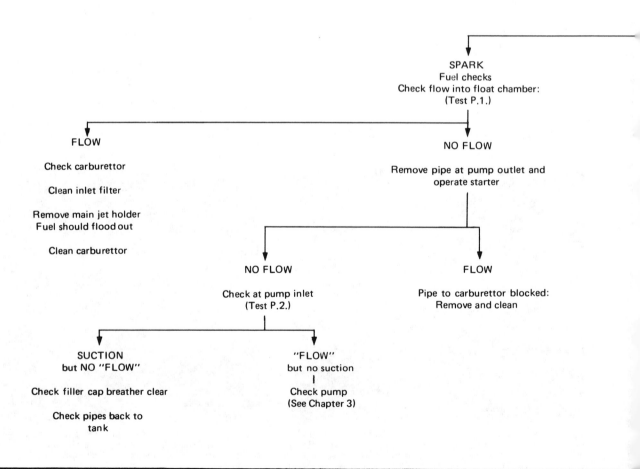

not run

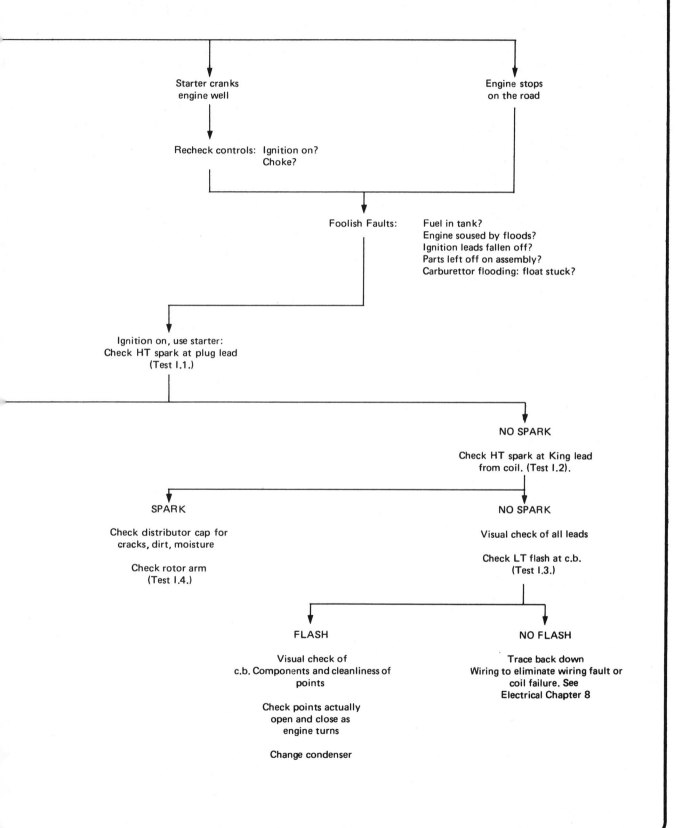

3 Fault-finding - engine runs erratically

1 Erratic running is nearly always a partial fuel blockage. It is therefore best first to eliminate any ignition failures.
2 An ignition fault that gives erratic running will probably be a loose lead. Anything else would give difficult starting. A check should therefore be made of all leads.
3 Having dismissed the ignition system, carefully note the circumstances that provoke the erratic running, and then refer to the fault section of Chapter 3.

4 Knocks and noises - roughness or smoke

1 The car will often give audible warning of mechanical failure in very good time. If these are heeded when faint and diagnosed then, disaster and more expensive repair bills can be avoided.
2 You will need to know how to interpret noises when you are buying a second hand car. If you are inexperienced then you will need help. A run in a similar car but one known to be in good mechanical order can set your standards.
3 Then as you get to know your car you will learn its normal noises and must be alert to the possibility of new ones appearing. Listening to what the car has to say is helped by ruthless tracking down of minor rattles.
4 There is a tendency to be rude about the noise a FIAT 500 makes. The overall noise produced is not very loud, but it is accentuated by the nearness of the engine and its air cooling cowling.
5 The general noise level is fairly continuous, and difficult to locate through. Noises due to defects are heard through this background, are usually not continuous, being provoked by some circumstance.
6 Rough running due to the partial failure of a cylinder is sometimes difficult to detect. The complete failure of one when there are only two is very obvious on this car! A smoky exhaust, or excessive oil consumption are important symptoms. Beware of mistaking tight wisps of vapour of condensation in cold weather as smoke.
7 Two tests can help to keep a check on a car's condition: These are acceleration and compression tests, and are described later.
8 Adjacent are tabled various defects. Each is treated individually. In practise faults or wear may, or probably will occur simultaneously. So neither the symptoms nor the faults would be so clear-cut. Tests referred to in the diagnosis tables are described with the fault-finding tests on later pages.
9 If the diagnosis tells you something serious is amiss it would be wise to get a second opinion. If you decide to get a reconditioned engine it is sufficient to learn that the old engine is badly worn. If you are going to overhaul the engine yourself, then a more exact diagnosis could help you decide whether the work really is within what you think you can cope with, and that it is not bad enough to warrant a reconditioned one. It would also let you order up the spares in advance. If the engine is in good enough order to continue to run without damaging itself and making the subsequent repair much more expensive, then there is opportunity to prolong the observation of the symptoms, so assess them better. Finally, the ultimate most accurate and thorough diagnosis is to take it apart and look inside.

5 Details of fault-finding tests

The systematic fault finding chart calls for various tests to be done. These are given below.
1 Tests of the ignition system:

Test 1.1.
Check ignition HT at a plug.
   a. Switch on
   b. Take lead off a plug
   c. Hold metal contact of the fitting on the end of the lead 1/8th inch from a bright metal "earth"
such as the cylinder head. If the plug lead fitting has a shroud to cover the plug stick a ¼ inch bolt into the contact as a probe.
   d. Operate the starter (by an assistant, or direct to the switch on the starter with your other hand).
   e. There should be an easily noticeable spark.

Test 1.2.
Ignition HT at source.
   a. If possible take the king lead from the distributor and hold the end 1/8 inch from earth and then check as for Test 1.
   b. If HT lead is not readily detachable from the distributor cap, remove it from the coil, and rig up a temporary lead.

Test 1.3.
Check ignition LT at contact breaker.
   a. Remove distributor cap
   b. Ignition switch on
   c. Open cb points with a thin screwdriver, or if points already open on the cam, short them with the screwdriver.
   d. There should be a small but definite spark.

Test 1.4.
Check the rotor arm.
   a. This test is to see if there is a short through the rotor arm's body to the spindle underneath.
   b. Rig up the king lead or a substitute as for Test 2.
   c. But hold the lead near the centre of the contact arm on the rotor
   d. Operate the starter
   e. There should be only the one small spark as the metallic mass of the rotor arm is electrically charged, and then no further sparks.
   f. Continued sparks mean there is current flow to somewhere; thus a faulty rotor arm.

2 Tests of the petrol system:

Test P.1.
Check fuel flow into the carburettor.
   a. Remove the large nut near the union for the pipe into the carburettor on the float chamber top, to expose the filter.
   b. Operate the starter
   c. Fuel should flood into the cavity from the pipe, and overflow once the float chamber is full.

Test P.2.
Check fuel flow into the pump.
   a. This is a difficult check unless the car is facing steeply uphill.
   b. In this case removal of the pipe into the pump from the tank should allow fuel flow by syphoning.
   c. If syphoning does not occur, try blowing back down the pipe. This should be an easy blow, and an assistant should be able to hear the air bubbling out in the tank.
   d. The blow could even have unwittingly cleared a stoppage, and it is worth trying to start after it.
   e. Also, with the pipe off the pump inlet, and the engine turning on the starter, a finger over the inlet union should feel suction.

3 Test of engine acceleration:

It is very difficult to judge properly if the engine is giving its correct power. An objective test is needed.
   a. Choose a long straight hill up which the car can just accelerate in the speed range 35 to 45 mph.
   b. Choose prominant landmarks at beginning and end of the test stretch.

Chapter 1/Engine

Symptom	Test/Circumstance	Probable Cause
Poor acceleration, otherwise smooth and quiet	Compression test satisfactory	Ignition timing wrong. Brakes binding. Blocked air cleaner. Accelerator linkage maladjusted.
Poor acceleration: Oil consumption and smoke satisfactory	Compression test low	Burned exhaust valve.
Idle probably rough (continuously and rythmically irregular)		Tappets no clearance.
Oil consumption high, exhaust smoky; some loss of power	Compression test low	Worn pistons, rings and cylinders.
Oil consumption perhaps above normal. Exhaust smoky, particularly after long idle	Compression test satisfactory	Inlet valve stem oil seals incorrectly fitted. Oil filler valve stuck open.
Ditto, with some loss of power	Ditto	Worn valve guides.
Extreme loss of power. Engine idle very rough, on one cylinder like a motor-cycle	Pull off each spark plug lead and replace it in turn (wear thick glove)	Engine misfiring on cylinder whose plug makes no difference. Spark plug faulty. Broken rocker.
Ditto	Ditto, plus compression test which shows bad cylinder very low	Valve jammed. Piston broken.
Oil dipstick blows up out of hole in crankcase	None	Oil filler valve stuck shut.

Noise	Circumstance	Possible Cause
Engine faults in general	Slow down with engine speed, and disappear in neutral with the engine switched off.	
Light tapping	At all speeds and loads, though drowned by others at speed in top gear	If slight, tappets too wide. If bad, worn valve gear.
Continuous light clatter		Worn timing chain.
Loud hollow knock	Worst cold and under load	Piston slap. Ignore if engine otherwise good. Test compression.
Loud solid knock, with probably oil warning light coming on at idle, hot	At idle when hot. At speed hot. In extreme cases a loud and wild hammering at particular engine speeds. Disconnecting a spark plug alters knock	Big ends excessive clearance. In mild cases new shells will cure. If the engine used when medium bad crankshaft will be hammered. In bad cases the con rod breaks and wrecks engine.
Dull low thudding, and low oil pressure	Worst at speed, hot.	Worn main bearings.
Continuous whining or roaring noises	Dependant on engine speed, but alters in different gears, and when pulling or on over-run	Transmission wear or wrong meshing of final drive.
Continuous roaring	In neutral, engine off	Wheel bearings.
Continuous swishing and occasional clonks	Free-wheeling, engine off	Drive shaft splines at hubs dry and rusty, and shatter springs u/s.

For other gear grating noises and clonks see clutch, Chapter 5.

c. If possible time the car over the test stretch with a stop watch. Anyway, note the speedometer reading at the beginning and end.
d. Always enter the test stretch at the same speed. Only do tests in conditions of light winds.
e. Do the test a number of times when you know the car is going well. Also try and do it with another similar car, known to be in good order. Record these results so that when you are suspicious of the car's performance a test can straight away give useful information.

4 Test of engine compression:

Useful information in defect diagnosis is the amount of pressure that can be achieved in the cylinders. This will indicate the state of the pistons in relation to the bores, as they must build up the pressure, and also it shows whether the valves are sealing the cylinders properly.

a. An engine compression test gauge is needed. This is the sort of equipment the most enthusiastic owner-mechanic gets, but probably the FIAT 500 owner will have to get a garage to do this test.
b. Warm up the engine. Remove the spark plugs.
c. Hold the rubber seal of the gauge tightly over the spark plug hole in the cylinder head.
d. Get an assistant to operate the starter whilst holding the throttle open wide. The starter will need to work for about 3 seconds, to allow a reading to stabilise on the gauge.
e. Note down the reading.
f. Release the pressure from the gauge, and then do the test on the other cylinder. After that do both cylinders for the second time. If the second reading is more than 2% different from the first for that cylinder, do a third test to get an average.
g. The readings for the two cylinders should be within 5% of each other. The reading should be about 100-107 lbs in^2 (7 - 7.5 kg cm^2). But more important, it is the difference between the two that gives the indication of poor compression.

5 Stethoscope for engine noises:

A simple stethoscope can be made to listen to odd engine noises.

a. Get a piece of plastic petrol pipe about 3 ft long.
b. In one end put a probe of thin metal pipe about 4 inches long.
c. Put the end of the plastic tube in an ear, and search for noises with the metal end.
d. The stethoscope will probably not held in locating major knocks deep in the engine, such as big ends, but it is good for locating strange noises in such things as the dynamo.

Chapter 2 Lubrication, cooling, heating and exhaust

Contents

Lubrication description ... 1	Overcooling ... 7
Lubrication care ... 2	Thermostat ... 8
Lubrication problems ... 3	Interior heating/ventilation ... 9
Air cooling description ... 4	Heater safety device ... 10
Care of the cooling system ... 5	Exhaust ... 11
Overheating ... 6	

Specifications

Oil pressure: normal ...	36 - 43 lbf/in^2 (2.5 - 3 kg/cm^2)
Warning light comes on ...	14 - 8½ lbf/in^2 (1 - 0.6 kg/cm^2)
Thermostat: starts to open ...	70° - 80°C (81° - 85°C station wagon)
open wide...	81° - 87°C (91° - 97°C station wagon)

1 Lubrication: General description

1 The lubrication system has been separated from other engine details to highlight the fact that in addition to limiting wear, oil flow in the lubrication system does a lot of cooling work, and particularly on this engine as it is air-cooled.

2 The oil is circulated by a gear pump driven by the rear end of the camshaft. On the normal, upright, engine the pump is directly on the end of the camshaft. The horizontal engine of the station wagon has its pump at the bottom of a vertical shaft which is driven by a gear wheel on the end of the camshaft. In both cases the pump is in the timing chain cover. To relieve excess pressure, particularly when cold, there is a pressure relief valve (PRV). The PRV is on the rear end of the camshaft. Loss of pressure will be signalled by a sender screwed through the side of the crankcase into the main oil gallery and wired to a warning light on the dashboard. The warning light comes on with the ignition, so before the engine starts and pumps up pressure it lights at the same time giving a chance to test it. It should come on if the pressure falls to 14 - 8½ lbf/in2 (1 - 0.6 kg/cm2).

3 There are two filters. A strainer on the end of the pick-up pipe in the sump takes out large particles. Fine ones are separated centrifugally, being flung outwards as the oil passes through the belt pulley on the rear end of the crankshaft.

4 Oil cooling is done in the sump. There is a special double skin which guides cold air direct from the fan.

5 Condensation and piston blow-by gasses are sucked out of the crankcase by a breather system. These are piped to the carburettor, so that they are burned by the engine.

2 Care of the lubrication system

1 Buy top quality oil. Oil must lubricate well both when hot and cold. It must neutralise contaminating by-products of combustion, particularly those formed in the period of running after a cold start. The oil must act as a detergent to keep the sludge of by-products in suspension, lest they lacquer on to moving parts, or block oil ways. The oil must prevent corrosion when the engine is not running.

2 The oil needs changing at least as often as in the maintenance schedules: more frequently rather than less.

3 As part of the 12000 mile task the centrifugal filter is cleaned. The work is described in Routine Maintenance.

4 A good oil, particularly on a car not just used on short journeys, should keep the engine clean inside, but whenever any part is stripped the opportunity should be taken to clean it inside.

5 Overhaul of the oil pump was detailed in Chapter 1:21.

3 Lubrication problems

1 The oil level must be kept correct, to the marks on the dipstick. If too low there will be loss of oil pressure on corners, which would damage bearings. Too high a level causes excessive oil to be thrashed about by the crankshaft.

2 If the oil pressure warning light suddenly comes on, put the gearbox into neutral at once, and free-wheel to park off the road. The switch could be at fault, but if a real alarm the engine

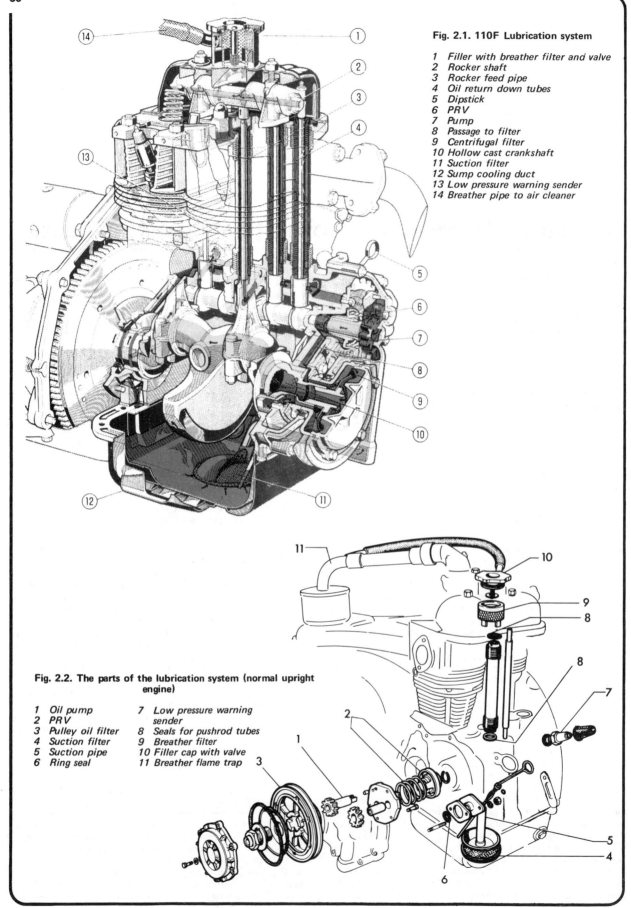

Fig. 2.1. 110F Lubrication system

1. Filler with breather filter and valve
2. Rocker shaft
3. Rocker feed pipe
4. Oil return down tubes
5. Dipstick
6. PRV
7. Pump
8. Passage to filter
9. Centrifugal filter
10. Hollow cast crankshaft
11. Suction filter
12. Sump cooling duct
13. Low pressure warning sender
14. Breather pipe to air cleaner

Fig. 2.2. The parts of the lubrication system (normal upright engine)

1. Oil pump
2. PRV
3. Pulley oil filter
4. Suction filter
5. Suction pipe
6. Ring seal
7. Low pressure warning sender
8. Seals for pushrod tubes
9. Breather filter
10. Filler cap with valve
11. Breather flame trap

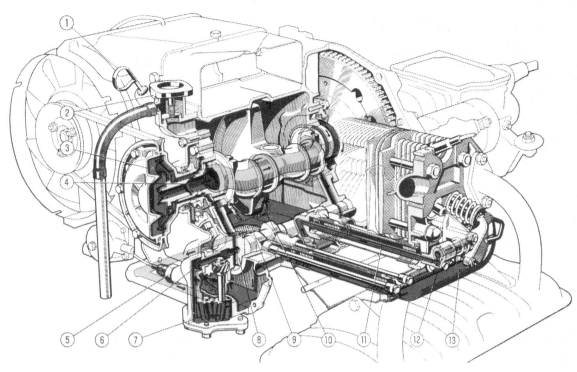

Fig. 2.3. 120 Lubrication system

1 Dipstick
2 Oil filler with breather pipe (early type open to atmosphere)
3 Centrifugal oil filter
4 Crankshaft, cast, hollow
5 Low pressure warning sender
6 PRV
7 Pump
8 Oil gallery along camshaft
9 Oil suction filter
10 Sump drain plug
11 Pipe to rockers
12 Rocker shaft
13 Rocker cover

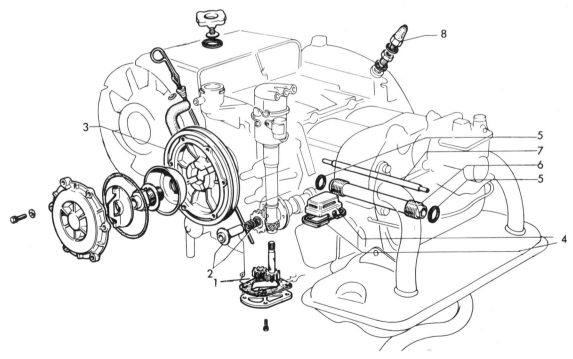

Fig. 2.4. Lubrication system parts of the horizontal engine

1 Oil pump
2 PRV
3 Oil filter
4 Oil suction filter
5 Seals, for
6 Tube, for
7 Push rod
8 Low pressure warning sender

could be wrecked in a few yards. Loss of oil pressure is normally accompanied by the clatter of unlubricated bearings, as oil is no longer filling their clearance. If the engine sounds healthy, try a new sender unit: but it would be unwise to drive the car to a garage to get it: bring it to the car.

3   With a worn engine loose bearings allow oil to leak out, so oil pressure is low particularly when the oil is hot and thin, and at idle when the pump outlet low. This will show itself by the flashing of the warning light. How dire the situation is, depends on the specific setting of the individual pressure warning sender. But its indication forms one of the symptoms in the diagnosis of a worn engine.

4   Once you become interested in engine diagnosis you will want to fit an oil pressure gauge. This shows what is going on in greater detail. The warning light is still useful to attract attention in the event of sudden failure. The oil pressure gauge is plumbed into the engine by inserting a 'T' union into the tapping for the sender unit. It is vital that the fitting is well done, lest oil be lost and the engine wrecked. Normal oil pressure is 36 - 43 lbs/in2 (2.5 - 3 kgkm2).

5   Ensure the air passages around the sump are kept clear of restriction by dirt and are not damaged.

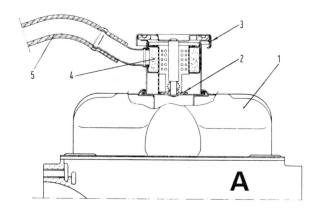

4 Air cooling - description

1   The air is blown over the engine by a fan on the dynamo shaft. The air is guided over the engine by a cowling which envelopes the whole.

2   The air is drawn in through a grille in the body, and passes over the large fins on the cylinders and the head, which therefore have a large surface area from which to dissipate the heat. At the bottom a special scoop guides part of the air down to the sump.

3   The air that goes by the sump then leaves the system. The air that goes over the cylinders has two possible routes. These are controlled by the thermostat and the heater control.

4   When the engine is cold the air flow needs to be restricted or it will be overcooled. The thermostat is a bellows that expands when hot. This movement is used to control the air outlet valve. The thermostat holds this shut till the engine warms up, then progressively opens to allow cooling when needed. Another branch of the outlet channel leads to the interior heater, beside that leading past the thermostat. When heat is needed inside the car the air can be let in by opening the valve on the heater channel on the floor in the back. It might be thought that use of the heater when the engine is still cold could allow overcooling, as the passage is not controlled by the thermostat, but the passengers should only turn on the heater when it is giving out hot air: also the air coming down the heater channel is restricted by its length and narrowness.

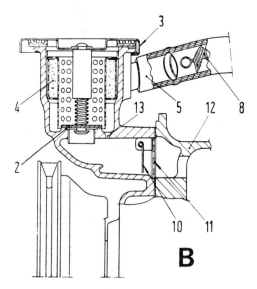

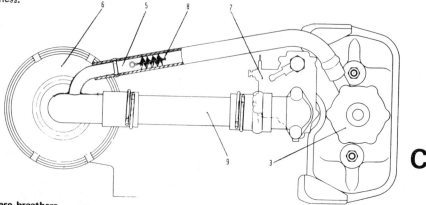

Fig. 2.5. Crankcase breathers
A. Cross section sedan. B. Cross section station wagon C. Plan view sedan

1 Rocker cover	4 Breather filter	7 Carburettor
2 Breather valve in	5 Breather pipe	8 Flame trap
3 Oil filler cap	6 Air cleaner	9 Air intake pipe
		10 Station wagon liquid seperator and,
		11 Filter, and
		12 Crankcase
		13 Byepass

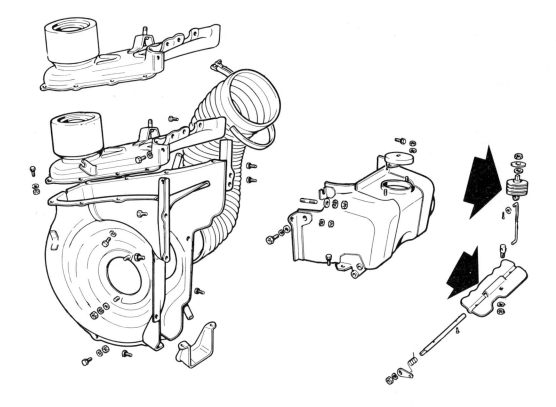

Fig. 2.6. Air cooling cowling for the normal type 110 engine, the thermostat and its flap valve being arrowed

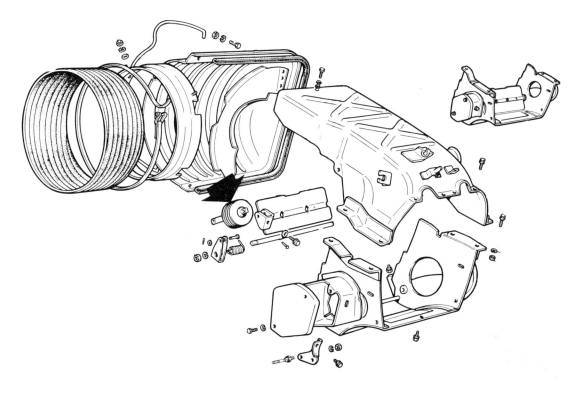

Fig. 2.7. Cowlings for the 120 engine of the station wagon. The thermostat is arrowed

5 Care of the air cooling system

1 The cooling system depends heavily on the fan belt. Checking the tension of this is part of the 3000 mile maintenance task, and it is replaced as preventative maintenance to avoid failure on the road at 18,000 miles. The work was described in Routine Maintenance.

2 Though it is difficult to do anything about it, try to avoid getting anything sucked into the air trunking that will build up on the fins on the engine. For example when parked under a tree in the autumn quite a number of leaves can get down into the air intake.

3 Check the cowling for security or cracks when doing the 3,000 mile task inspection of the car. Cracks should be arrested by drilling a hole just beyond their end, or else they will propagate further. Drill a ¼ in hole. Then reinforce the metal either side of the crack by glueing over it a strip of steel about 1 - 1½ in wide. Stick it with a strong adhesive such as Araldite.

6 Overheating

1 Symptoms of overheating are often difficult to detect. One instant symptom is that heater output may be low, as air is not coming through properly, and not taking away heat. If an oil pressure gauge is fitted the pressure will be low as the oil will be thinner than normal.

2 Sometimes the engine will show symptoms in good time. In hot weather fuel vapour locks will give engine stoppages. In any weather pinking may be provoked. (Pinking is bad combustion inside the cylinder under load, and sounds like muffled bells in the engine). In worse cases the engine may loose power and get noisier; what can only be described as sounding distressed. In the end parts get so hot and mis-shaped, and the oil so thin, that the engine seizes. Should this disaster occur, apart from finding and curing the cause of overheating the attempt should be made to save the engine, as follows. As soon as the engine seizes, or goes stiff, loosing power, if the seizure is not total, freewheel to a

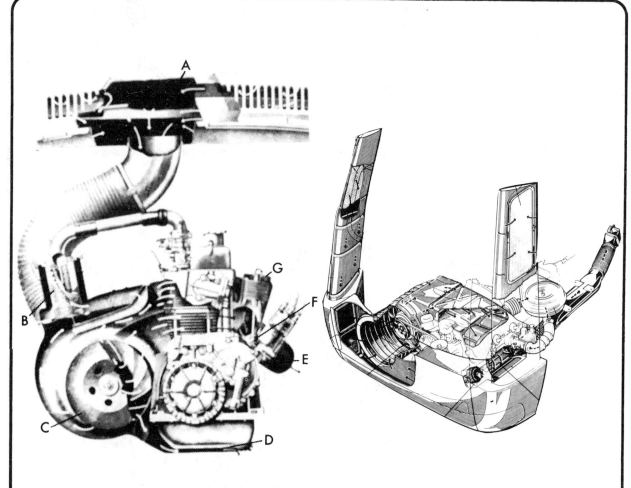

Fig. 2.8. Air circulation, type 110.

A Air intake
B Engine air cleaner
C Centrifugal fan
D Sump cooling duct
E Warm air to heater
F Outlet flap valve
G Thermostat

Fig. 2.9. Cooling circuit for station wagon, type 120.000

stop, getting off the road, and allow the engine to cool. Remove the spark plugs, and squirt in some engine oil. Turn the engine over a few times by hand. Then try running the engine gently. Treat it as a new one. The bores will have been scuffed by the seizure; the engine will need running in. Initially there may be odd noises as the piston rings ride over the scuff marks. If these do not dissapear after 20 miles the engine would appear to need stripping to examine the damage.

3 Overheating can be caused by things other than failure of the cooling system. The ignition timing a long way out can cause it. Also a very weak fuel air mixture due to a carburettor defect, or an air leak into the inlet beneath the carburettor.

4 If the engine overheats make a visual check. Check the fan belt. Look at the thermostat, or at least the flap valve it operates. If the engine has just been running apparently hot, then this should be open. On the sedan it starts to open at 70° - 80° and is fully open at 81° - 87° C. On the station wagon the thermostat starts to open the flap at 81° - 85° and is wide open at 91° - 97° C.

5 If no external fault or failure can explain the overheating, then the whole cowling will have to be removed, as probably dirt is blocking air flow, or the fins on the engine cannot give up their heat to the air due to layers of grime. The removal of the cowling can be done, with difficulty, with the engine in place, and was described in Chapter 1:6/2.

Fig. 2.10. The anchor plate in the right half of the cowling which holds up the thermostat

7 Overcooling

1 If the flap valve jams open, the engine will run excessively cool, and this is bad for it. It would be worst on a car used for short journeys. The oil will be too thick, there will be a lot of condensation inside the engine, and not enough heat to drive it out. The mechanical fit of pistons and cylinders will be wrong, as the aluminium pistons shring more than the cast iron cylinders from the size at which they normally run when hot. The oil will be too thick.

2 In this case the flap valve will be seen open. This should be looked at as a matter of routine during checks of the engine compartment. Another symptom of overcooling will be cool air coming from the heater. If an oil pressure gauge is fitted, the pressure will be seen to be higher than normal.

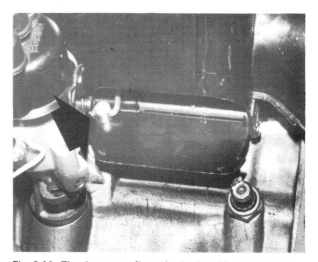

Fig. 2.11. The thermostat flap valve in the wide open position. The bottom end of the actuating rod can be seen where it bends into its fixture on the valve

8 Thermostat

1 The important parts this plays in the cooling system has been described. By noting the position the thermostat puts the flap valve into after various short or long journeys when the car is healthy gives the knowledge to be able to judge when it is at fault.

2 When it fails, the thermostat will normally leave the flap valve in the open position.

3 The operation of the thermostat can be checked by immersing it in a saucepan of boiling water. To remove the thermostat from the engine to do this, undo the two nuts on the studs holding the plate where the thermostat is anchored to the cowling. Then undo the fixture for the operating rod that connects the thermostat to the valve by undoing the nut on the outside. To reassemble it is rather a struggle, as there will be the tension of the thermostat in the shut position.

9 Interior heater/ventilator

1 Air from the cooling system is blown along the spine of the car to the front.

2 There is a master control lever at the rear. Then at knee level at the front are the distribution controls. If these are shut more air goes to the windscreen for demisting.

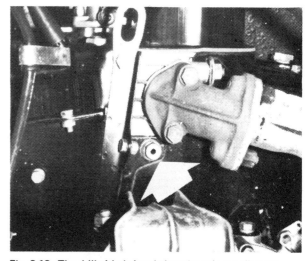

Fig. 2.12. The drilled bolt just below the exhaust elbow to allow gas leaking past the head gasket to escape lest it be taken by the heater into the car

10 Heater safety device

1 If the cylinder head gasket is blowing, the escaping gases will get into the cooling air stream, and from there to the heater, and so poison the driver and passengers.

2 On the 110 F engine there is a special safety duct to lead such fumes away. A groove is cut in the top face of the cylinder, and a corresponding duct in the head. The latter connects to holes at front and rear of the head. In these holes are bolts with holes drilled down their centre. These bolts also secure the cowling. The leaking gases can therefore escape down the bolts and to the outside of the cowling.

3 A car fitted with this safety device can be easily recognised, as the drilled bolt can be seen below the exhaust outlet elbow. On cars too old to have it, attention must be paid to check for the smell of fumes when the heater is turned on, as the poisoning from them can creep on insidiously, and can be fatal.

4 For cars with the safety device a check can be made if a leaking gasket is suspected. Listen with a stethoscope as described at the end of the fault finding in Chapter 1. This will allow the differentiation to be made between a gasket leak and one from the exhaust flange on the head.

11 Exhaust

1 There is a down pipe from each end of the cylinder head: on Sedans there is an elbow between the head and the pipe; on station wagons it is fitted direct.

2 The removal of the complete exhaust unit is simple. The bolts are removed from the head, and then the silencer bracket from the crankcase.

3 When refitting the exhaust all joints must be put up finger tight before any one is tightened to allow the parts to settle into place without distortion. Use new gaskets, lightly smeared with grease at the flanges. Check that the unit is in place, and not strained. When the nuts and bolts are finger tight, then tighten the pipe flange to the elbow (not applicable to the station wagon), then the bracket holding the silencer to the crankcase. Then finally tighten the flange to the head. After the engine has been run retighten all joints.

Fig. 2.13. When tightening the exhaust, the flange joining the pipe to the elbow should be tightened first, and the elbow to the head last

Fig. 2.14. The silencer with its bracket is held to the crankcase by four nuts

Chapter 3 Fuel system

Contents

Introduction and description ... 1	Checking and testing the pump ... 7
Removal of carburettor ... 2	Water in the fuel ... 8
Stripping the carburettor ... 3	Cable controls ... 9
Idle adjustment ... 4	Fuel pump defects ... 10
Removal of petrol pump ... 5	Fault finding ... 11
Stripping the petrol pump ... 6	

Specifications

The following Weber carburettor types have been fitted:

New 500	26 IMB 1
500 Sport	26 IMB 3
500 D	26 IMB 4
Station wagon	26 OC
500 F	26 IMB 6
500 L	26 IMB 10 (incorporates CO limiter for idling)

Jets and chokes

	26 IMB 1 - 26 IMB 4 26 IMB 6 - 26 IMB 10	26 IMB 3	26 OC
Throat diameter ...	(26 mm)	(26 mm)	(26 mm)
Venturi ...	(21 mm)	(22 mm)	(20 mm)
Main jet diameter ...	(1.12 mm)	(1.25 mm)	(1.05 mm)
Idle jet diameter ...	(0.45 mm)	(0.45 mm)	(0.45 mm)
Starting jet diameter...	(0.90 mm F5)	(0.90 mm F5)	(0.80 mm F3)
Main air jet diameter ..	(2.35 mm)	(2.35 mm)	(2.10 mm)
Needle valve seat diameter ...	(1.25 mm)	(1.25 mm)	(1.25 mm)

Petrol pump ... Engine driven diaphragm type with filter, by Weber, BCD, or Savara.

1 Introduction and description

1 The fuel from the tank at the front of the car is fed by an engine driven pump to a downdraught carburettor (side draught on the station wagon).
2 The carburettor is of Weber manufacture, incorporating a manual two position starter device for cold starts. This is a separate section of the carburettor, and does not incorporate a "strangler", (the misnomer of "choke" is often applied to it).
3 The pump is driven from an eccentric on the camshaft by a long rod across the engine. At least five models of pump from various manufacturers have been used on this car.
4 The carburettor and fuel pump normally only receive attention when giving trouble. This is for once correct: by ensuring dirt never gets in the system, it should not need attention for long periods, so need not be exposed to the greater risk of dirt entering when dismantled.
5 When problems occur they should be tackled as follows. First refer to the general fault-finding chart at the end of Chapter 1, to see if the trouble seems to be in the fuel system. If it seems to be so, then refer to the detailed fault tabulation at the end of this one. Having established the likely cause, strip the suspected component as described in the following sections.
6 Due to the multiplicity of small holes, corners, and narrow passages, dirt is the biggest enemy of these carburettors. Dirt gradually builds up in pockets. If it is disturbed, it moves down stream, in a lump, and causes a blockage elsewhere.
7 There are separate passages for idling, full power, etc: partial blockage of only some can give peculiar results.

2 Removal of the carburettor

1 There comes a time when the carburettor needs a complete clean-out. Then it must be done very thoroughly and scrupulously; so it is best taken off the engine for it.
2 Remove the air cleaner connection at the carburettor.
3 Disconnect the petrol pipe at the inlet to the carburettor.
4 Disconnect the choke inner and outer cables at the carburettor.
5 Unclip the throttle linkage to the carburettor at the relay lever on the air cowling.
6 Take off the rocker box.

7 Undo the two nuts fixing the carburettor to the top of the cylinder head, and lift it off with the drip tray.
8 Cover the inlet to the engine to prevent anything falling in.
9 When reassembling use new gaskets between the engine and the drip tray, and between tray and carburettor. Check for leaks after starting up.

3 Stripping the carburettor

1 Having removed the carburettor, clean its outside thoroughly.
2 Then take off the bolt covering the filter in the float chamber top, and lift on the filter (photo).
3 Take out the screws holding on the carburettor top (photo).
4 Remove the top, taking care of the float swinging about on its pivot in the top, (photo).
5 Remove the pin holding the float, and taking care the needle valve is not dropped, take off the float and lift out the needle valve, (photos).
6 Unscrew the main and idle jet holders from outside the body, (photos).
7 Remove the choke jet and air correction emulsion tube from inside, (photo).
8 Take the choke/cold starting device off the side, (photo).
9 Clean the interior of the carburettor, and all the parts removed. Only blow through the jets. Never use anything metal to clean them, such as poking with a piece of wire, or their size will be altered. Do not let dislodged dirt get down any of the passages.
10 Reassemble all the parts, ensuring no wisps of rag are in any of the passages. If any do get in, dip the jet in petrol and light with a match. Check that the float works the needle valve smoothly, its small lever being at right angles to the axis of the valve.
11 If the float is suspect, shake it. If it is punctured, little drops of petrol will be heard pinging about inside. To confirm the puncture, immerse the float in boiling water and watch for bubbles rising from the hole. A punctured float must be replaced with a new one. Do not try to repair.

3.2. Take off the bolt and lift out the filter

3.3. Undo the screws holding the top

3.4. Lift off, taking care not to damage the float

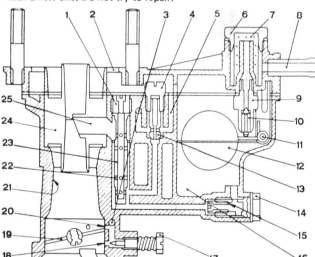

Fig. 3.1. Section of Weber 26 IMB 4 carburettor, of the sedan

1 Air corrector jet
2 Air inlet
3 Idle mixture duct
4 Idle jet holder
5 Idle air orifice
6 Filter cover
7 Filter
8 Fuel inlet
9 Needle valve seat
10 Needle
11 Float pivot
12 Float
13 Idle jet
14 Main jet holder
15 Main jet
16 Float chamber
17 Idle mixture adjustment screw
18 Idle mixture orifice
19 Throttle
20 Transition hole
21 Primary Venturi (not interchangeable)
22 Emulsion orifices
23 Emulsion well
24 Secondary Venturi (not interchangeable)
25 Main nozzle

3.5a. Pull out the pin

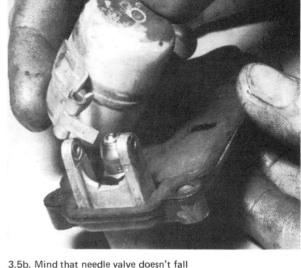

3.5b. Mind that needle valve doesn't fall out as the float comes away

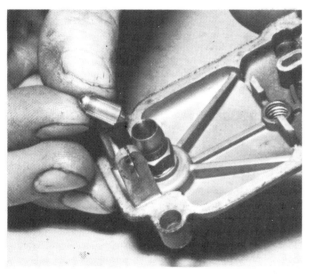

3.5c. The needle valve

3.6a. Take out the idle jet,

3.6b. and the main jet holder

3.6c. If there is dirt behind the jet, take it out of its holder with a screwdriver

3.7. Take out the emulsion tube (black arrow) and starting jet (white arrow)

3.8. Take off the starting device

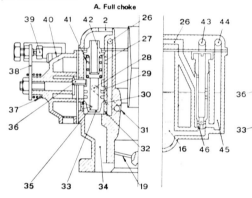

Fig. 3.2. Sections of 26 IMB 4 Weber carburettor starting device

- 2 Air inlet
- 16 Float chamber
- 19 Throttle
- 21 Primary venturi
- 24 Secondary venturi
- 26 Mixture duct
- 27 Mixture weakening air orifice
- 28 Transition duct
- 29 Transition mixture orifice
- 30 Starting mixture orifice
- 31 Transition orifice
- 32 Starting mixture orifice
- 33 Starting valve
- 34 Mixture duct
- 35 Starting device air orifices
- 36 Rocker
- 37 Lever return spring
- 38 Starting device control lever
- 39 Control wire screw
- 40 Cover with support for starting device control cable
- 41 Starting valve spring
- 42 Spring casing
- 43 Starting jet emulsion air orifice
- 44 Air emulsion reserve well orifice
- 45 Starting reserve well
- 46 Starting jet

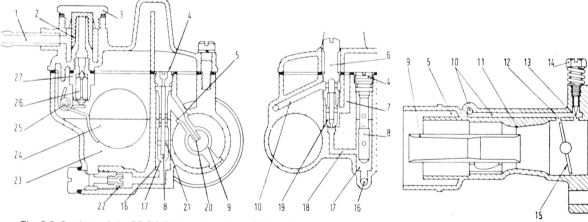

Fig. 3.3. Sections of the 26 OC Weber carburettor of the station wagon

- 1 Fuel inlet
- 2 Filter gauze
- 3 Filter plug
- 4 Air corrector jet
- 5 Air intake
- 6 Idle jet holder
- 7 Idle air duct
- 8 Emulsion tube
- 9 Secondary venturi
- 10 Idle mixture duct
- 11 Primary venturi
- 12 Progressive hole
- 13 Idle orifice to duct
- 14 Idle mixture adjustment screw
- 15 Throttle
- 16 Chamber-to-well duct
- 17 Emulsion tube housing well
- 18 Well-to-idle jet duct
- 19 Idle speed jet
- 20 Nozzle
- 21 Emulsion orifices
- 22 Main jet
- 23 Float chamber
- 24 Float
- 25 Float pivot
- 26 Valve needle
- 27 Needle valve

Chapter 3/Fuel system

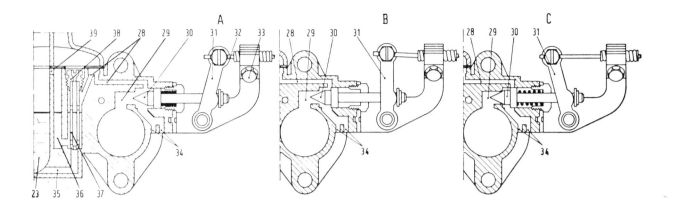

Fig. 3.4. The 26 OC Starting device

23 Float chamber
28 Starting mixture duct
29 Starting mixture part
30 Starting valve
31 Starting device control lever
32 Starting device control wire
33 Cable fixing screw
34 Emulsion air orifices
35 Bowl-to-starting jet duct
36 Starting reserve well
37 Starting jet
38 Starting air corrector screw
39 Reserve well emulsion air slot

A Full choke   B Partial choke   C Normal running

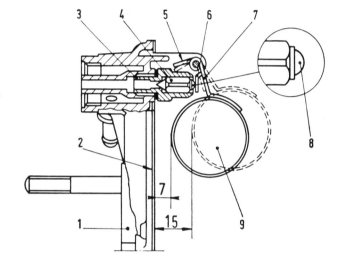

Fig. 3.5. Float level setting for sedan carburettor type 26 IMB 4

1 Carburettor cover
2 Gasket
3 Needle valve
4 Needle valve
5 Lug
6 Arms
7 Arms
8 Ball of the valve
9 Float

7 and 15 mm are the distances to the float in the two extremes of its movement.

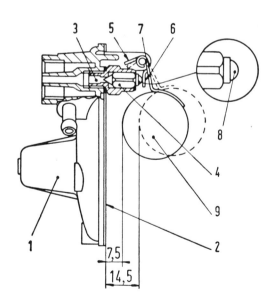

Fig. 3.6. Float level setting for the station wagon carburettor

1 Cover
2 Gasket
3 Needle valve
4 Needle valve
5 Lug
6 Arms
7 Arms
8 Ball of the valve
9 Float

7.5 and 14.5 mm are the dimensions of the float from the cover in the two extremes of its movement.

4 Idle adjustment

1 If the carburettor has been dismantled the idle will be in need of readjustment. It may also be found wrong after some period of use of the car, so is part of the 6,000 mile task.
2 To readjust the idle, first warm up the engine.
3 Then screw the throttle stop screw so that the engine idles slowly. As a guide, the ignition warning light should be just coming on.
4 Now adjust the mixture screw to give the fastest, smoothest idle. Having done this, the throttle stop may have to be screwed out a bit to slow down the idle again. The mixture screw is item 17 in Fig. 3.1. It is the screw with a coil spring to lock it which screws into the carburettor body.

5 Removal and replacement of the petrol pump

1 Disconnect the petrol pipe at the tank to prevent syphoning.
2 Disconnect the inlet and outlet pipes from the pump.
3 Undo the two nuts holding the pump to the crankcase.
4 Lift off the pump, gasket, plastic distance piece, and second gasket.
5 Refit using new gaskets, which should be lightly smeared with grease.
6 After the engine has been run, check for leaks, and retighten the nuts.
7 If difficulty is experienced fitting new petrol pipes, hold them tightly in the hand; the warmth will soften them.

6 Stripping the petrol pump

1 On pumps of the Weber and BCD types take off the domed cover and remove the disc filter. On the Savara type undo the inlet union from the body and take out its tubular filter.
2 Note the relative positions of the two halves of the pump body and mark them with a scratch. Undo the screws holding them together.
3 Part the two halves of the pump carefully, particularly if you have no spare diaphragms, lest it tears.
4 Wash the filter, valves, and body in clean petrol.

7 Checking and testing the pump

1 Examine the diaphragms for tears or cracks. Note the double layout of two diaphragms and spacer. The bottom one prevents petrol leaking into the engine.
2 Check the valves for wear or damage or trapped dirt.
3 Check the condition of the bowl gasket.
4 Reassemble the pump, tightening the screws holding the two halves together evenly, diagonally, and gradually.
5 Test the pump before refitting by working its lever by hand. Put a finger over first the inlet, and then the outlet. Suction should be developed at the inlet, held by the valves, and the inrush of air felt and heard as the finger is removed. At the outlet, pressure should build up after two or three strokes, and the valves should hold it for some ten seconds.

8 Water in the fuel

1 Water means trouble if it gets into the fuel system; not just because the car will not run on it; it is usually only present in small quantities. But its surface tension - its ability to hang together as a drop - is high, so it can sit in a jet and block it! It is heavier than petrol. Once in the carburettor the bowl layout of the float chamber and the passages to jets allow it to collect in the bottom of the carburettor in amounts large enough to stop the engine.
2 Conversely, in the tank, the pipe drawing the fuel is positioned a little way up from the bottom. So if water in the tank is only a small amount, it stays there.
3 Water can get into the tank from a number of sources. It will condense there from the damp in the atmosphere when the tank is only part full of fuel, so holding a lot of air. It can get in with fuel taken from an unreliable source. Though shielded well on this car by the boot lid, it can get in direct at the filler.
4 Once in the tank it will not evaporate out, as it is sealed in by the petrol floating on top. The only thing to do is to remove the tank and drain it. If water has got in, other foreign bodies are likely to be there too, so it should be given a good swill out.
5 If the car stops on the road with water in the carburettor try starting up with the air cleaner connection removed, and a hand over the intake to act as a choke. It may be possible to drag the water out of the carburettor. The carburettor itself should first be drained by removing the main jet holder. Then the engine must be turned over a few times on the starter to pump up fuel to refill the float chamber.

Fig. 3.7. Taking off the dome cover on a BCD or Weber pump

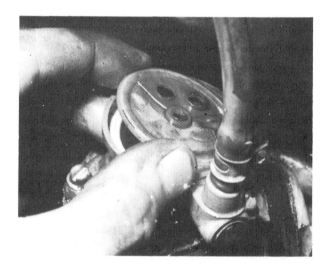

Fig. 3.8. The disc filter on the BCD pump

Fig. 3.9. Removing the pump

Chapter 3/Fuel system

Fig. 3.10. The Savara type pump without the removable domed cover. The filter is in the inlet union

Fig. 3.11. The fuel tank with the fitting for the sender for the gauge and low level warning, and the fuel feed pipe

Fig. 3.12. The front part of the heater trunking must be removed to get at the front end of the throttle cable. Lubricate the pedal pivots

9 Cable controls

1 When fitting new choke or accelerator cables the spine of the car floor must be opened up.
2 There is a cover at the rear end, on top, just in front of the rear seat.
3 At the front the ducting to the heater must be removed.
4 When fitting new cables, if the outer is plain ordinary steel, remove the inner, and coat it with grease of the molybdenum-disulphide type. This is not necessary if the outer cable is lined with plastic, and the inner stainless steel.
5 After connecting the cables, with an assistant in the driving seat, check as follows. The accelerator must give full throttle. When the pedal is released there must be some free play so that the throttle can go back onto its stop without being held from the idle position by the cable. The choke must have the full travel for the lever on the side of the carburettor.

10 Fuel pump defects

1 If the symptoms point to a fuel pump failure, then the problems are likely to be one of the following:
2 Blockage, by dirt, most probably of the filter.
3 Valve failure, either due to damage to a valve or its seat, or else dirt preventing it from seating properly. With modern clear plastic pipes this failure of the valves will show up. Under the action of the diaphragm, the fuel will be pumped up, but fall back again, because of the weak valves. It can be seen to go to and fro.
4 Diaphragm failure will give no sucking or blowing action. If punctured, petrol may flow out of a little hole in the spacer between the two layers of the diaphragm. (The lower layer is to stop petrol getting down into the sump).
5 An air leak will prevent the pump from sucking up from the tank. The most likely place for such a leak is the domed cover.

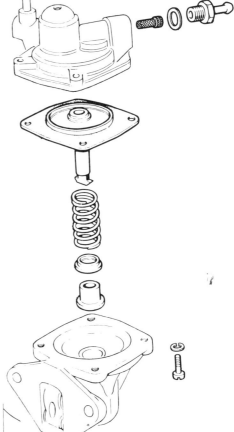

Fig. 13.13. The Savara pump

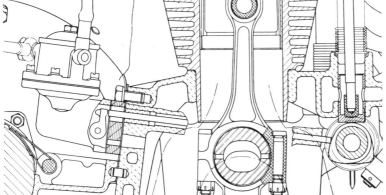

Fig. 3.14. Engine cross section through the fuel pump actuating rod

11 Fault finding

1 If the engine runs erratically, this can nearly always be blamed on the fuel system.
2 As some fuel is flowing it is very difficult to pin down just where is some partial blockage starving the engine. Therefore symptoms must be carefully noted, and reference made to the tables of faults on following pages.
3 Complete failure of the system is much easier to trace, and was dealt with on the fault finding chart at the end of Chapter 1.

Starting and running	Idling	Effect of choke	Probable defect
1 Starts well cold on choke. Only starts hot on wide throttle. Runs well at speed. Poor pick-up on part throttle	Will not idle	Does not help	Blocked idle jet and/or passages in carburettor.
2 Starts well hot or cold. Coughs and splutters when large throttle opening tried: Can hardly move	Idles well	Will help a bit to get car going	Blocked main jet.
3 Very difficult to start cold: Has to be 'warmed up on the starter'. Once warm goes perfectly	Normal	No effect	Blocked starter jet/passages or valve.
4 Engine tends to stall. Difficult to start hot. Goes well. High fuel consumption. Smell of petrol, and possibly petrol in drip tray under carburettor	Will not idle	Makes matters worse	Carburettor flooding:— Float jammed. Needle valve stuck or damaged or dirt on seating. Float set wrong: level too high on float chamber.
5 Varying symptoms. Sometimes cures itself. Sometimes goes in jerks, other times engine cuts out	Sometimes does, others not	Sometimes helps	Water in fuel.
6 Difficult to restart hot. In extreme hot weather may cut out when moving. Worse in mountains. Engine restarts when allowed to cool	Once restarted idles well	Makes little difference	Fuel vapour lock due to hot weather combining with engine overheating. Suspect fan belt loose; ignition timing wrong; obstruction to air flow.
7 Engine won't restart	Once restarted, behaves well	Makes matters worse	Plugs wet with petrol. Slowly press accelerator to floor: hold it there and work starter. (Not really a fault, just a temporary crisis).
8 Engine starts well hot or cold, then runs well initially, but peters out. Driving slowly may allow it to recover if symptoms mild	Initially idels well - will not idle after petering out unless symptoms mild	Will not rescue engine	Engine is initially running on the float chamber. There is a partial blockage between needle valve and tank. Perhaps filter, or a weak pump.
9 Ditto	Idles well	Helps a bit	Dirt getting sucked into main jet, but falling back out of it when flow stops. e.g. flake of dirt; wisp of rag. If latter, take out jet and burn it out.
10 Ditto, but damp cold weather. Cures itself if given a couple of minutes rest	Probably won't idle	Probably will not help	Carburettor icing. Check thermostat. Buy top quality but NOT high octane fuel.
11 Runs well up hill only. Worse when tank nearly empty	Peters out facing down hill	Nothing helps facing down hill	Fuel pump failure. Car is working on gravity feed.

Chapter 4 Ignition system

Contents

General description ... 1	Ignition advance - testing stroboscopic timing ... 7
Function ... 2	Condenser - removal and replacement... 8
Maintenance for distributor ... 3	General distributor wear ... 9
Contact breaker - adjustment ... 4	Sparking plugs ... 10
Ignition timing - static ... 5	Coil ... 11
Distributor - dismantling and refitting .. 6	Fault finding ... 12

Specifications

Distributor
Static advance ...	10° (equivalent to 13 mm round pulley)
Centrifugal: Sedan ...	18°
Station wagon ...	28°
Breaker contact pressure ...	16.8 ± 1.8 oz (475 ± 50 gr)
Contact gap ...	.019 inch to .021 inch (0.47 to 0.53 mm)
Breaker ...	78° ± 3°
Condenser capacity at 50 - 100 Hz ...	0.15 to 0.20 uF
Cam carrier shaft felt and oiler lubricant ...	Engine oil

Coil
Primary winding resistance at 68° ± 9°F (20° ± 5°C), not below	3.2 Ohms
Secondary winding resistance at 68° ± 9°F (20° ± 5°C) ...	5,000 ± 100 Ohms
Ground insulation resistance at 500 V d.c., not below ...	50 M Ohms

Spark plugs
Thread diameter and pitch ...	M 14 x 1.25
Type: Sedan ...	Marelli CW 225 N or 6 N
Station wagon ...	Marelli CW 260 N
All cars ...	Champion L 7 or L 87 Y
Point gap: Marelli ...	.020 inch to .024 inch (0.5 to 0.6 mm)
Champion ...	.024 inch to .028 inch (0.6 to 0.7 mm)

1 General description

1 The ignition is conventional coils with a distributor. On the sedan the distributor is on the right side of the engine. On the station wagon it sits on top of the shaft driving the oil pump, on the rear of the engine.
2 The coil is mounted on the engine compartment bulkhead, on the left on early sedans, to the right on later ones. On station wagons it is central, just in front of the engine.
3 The distributor mounting incorporates a slot for adjusting the timing.
4 Ignition advance is given by the centrifugal mechanism within the distributor.
5 The contact breakers are the same on the station wagon and sedan.

2 Function

1 For the engine to run an electrical spark ignites the fuel/air charge in the combustion chamber at exactly the right moment in relation to engine speed. The ignition system is based on supplying low tension voltage from the battery to the ignition coil where it is converted to high tension voltage. The high tension voltage is powerful enough to jump the sparking plug gap in the cylinders under high compression pressure, providing that the ignition system is in good working order and that all adjustments are correct.
2 The ignition system comprises two individual circuits known as the low tension circuit and the high tension circuit.
3 The low tension circuit (sometimes known as the primary circuit) comprises the battery, the lead to the ignition switch,

then to the low tension or primary coil windings, and the lead from the low tension coil windings, to the contact breaker points and condenser in the distributor.

4 The high tension circuit (sometimes known as the secondary circuit) comprises the high tension or secondary coil winding, the heavily insulated ignition lead from the centre of the coil to the centre of the distributor cap, the rotor arm, the sparking plug leads and the sparking plugs.

5 The complete ignition system operation is as follows. Low tension voltage from the car battery is changed within the ignition coil to high tension voltage by the opening of the contact breaker points in the low tension circuit. High tension voltage is then fed via the carbon brush in the centre of the distributor cap to the rotor arm of the distributor. The rotor arm revolves inside the distributor cap, and it comes in line with one of the metal segments in the cap, these being connected to the sparking plug leads. The opening of the contact breaker points causes the high tension voltage to build up, jump the gap from the rotor arm to the appropriate metal segment and so via the sparking plug lead to the sparking plug where it finally jumps the gap between the two spark plug electrodes, one being connected to earth: the engine.

6 The ignition time is advanced automatically to ensure the spark occurs at just the right instant for the particular prevailing engine speed, so that the flame can have burned fully when the piston starts to descend.

7 The ignition advance is controlled mechanically. The mechanical governor mechanism comprises two lead weights, which move out under centrifugal force from the central distributor shaft as the engine speed rises. As they move outwards they rotate the cams relative to the distributor shaft, and so advance the spark. The weights are held in position by two light springs and it is the tension of these that controls the amount of advance.

3 Maintenance for distributor

1 The distributor is serviced and lubricated as part of the 6,000 miles task.

2 Release the two clips securing the distributor cap to the distributor body, and lift away the cap. Clean the inside and outside of the cap with a dry cloth. It is unlikely that the metal segments will be badly burned or scored, but if they are the cap must be renewed. If a small deposit is on the segments it may be scraped away using a small screwdriver, (photo).

3 Push in the carbon brush located in the top of the cap several times to ensure that it moves freely. The brush should protrude by at least a quarter of an inch. If there are any cracks in the cap it must be replaced.

4 Take off the rotor arm by pulling it upwards.

5 If the contact breaker points are burned, pitted or badly worn, they must be removed and either replaced, or their faces must be rubbed smooth. It is a safe assumption that this will have to be done every 6,000 miles. It can also be reckoned that the points can be refaced once, but the next time must be replaced by a new set.

6 To remove the points prise off the little circlip on the moving contact's pivot post, and take off the washer underneath. Slacken the nut on the terminal bolt in the side of the distributor. Slide the spring contact blade off the moving contact from between the head of the terminal bolt and the insulator. Simultaneously slide the contact itself up its pivot, and remove it. Undo the screw holding the stationary contact to the distributor plate, and remove it too, (photos).

7 To reface the points, rub them on a fine carborundum stone, or on fine emery paper. It is important that the faces are rubbed flat and parallel to each other so that there will be complete face to face contact when the points are closed. One of the points will be pitted and the other will have deposits on it.

8 It is necessary to remove completely the built-up deposits, but not necessary to rub the pitted point right to the stage where all the pitting has disappeared, though obviously if this is done it

3.2. The distributor cap undone and rotor arm removed. (This one has the later type of oiler)

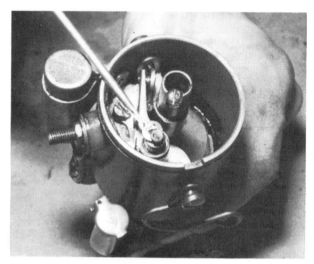
3.6a. Prise off the circlip on the moving contact post. (This distributor has the early type of oiler)

3.6b. Slacken the terminal nut, and slide the spring blade from under the bolt head, and the moving contact off the post

Chapter 4/Ignition system

will prolong the time before the operation of refacing the points has to be repeated.

9 Whilst the points are out of the way turn the engine over by hand to inspect both springs of the centrifugal advance mechanism underneath. The cam on the top of the distributor spindle should turn smoothly against the spring pressure. Lubricate with engine oil the advance mechanism, the wick on the top of the spindle, and the oiler on the side of the body (this is fitted with a felt wick). Smear a little grease on the cam.

10 Reassemble in reverse order.

11 Note the spring blade of the moving contact must go between the head of the terminal bolt and the sheet of insulator. As the nut on the terminal is tightened ensure the cut-a-way part of the large plastic insulator fits onto the hole in the distributor to hold the bolt central.

12 Put a drop of oil on the top of the post for the moving contact.

13 Now set the contact gap and retime the ignition as described in the next two sections.

4 Contact breaker - adjustment

1 The heel of the moving contact breaker rubbing on the rotating cam wears, so reducing the gap. If the gap is too small sparks may arc across, so no HT voltage will be built up for an HT spark. If too large the contact breaker will stay shut too short a time for the magnetic field to build up in the coil.

2 To set the gap turn the engine over by hand, using a 10 mm spanner on the bolts on the crankshaft pulley, till the cam in the distributor has opened the points as far as they will go, the heel of the contact being on the top of the cam.

3 Slacken the screw clamping the stationary contact to the distributor base plate. Put a .020 in (0.5 mm) feeler between the points. Move the stationary contact against the grip of the partially slackened clamping screw by inserting a screwdriver between it and the base plate. Move the contact so that the feeler is tightly held. Carefully tighten the clamping screw, without disturbing the points. Check the gap is correct, trying feelers of the size next, smaller and bigger, (photo).

4 Once the contacts have been used a lump builds up on one contact, making it impractical to check the gap unless the points are resurfaced. So when setting them check carefully that the gap is correct, and if any error does creep in it is better to have the gap a thousandth of an inch too large rather than too small. Then it will last to the next 6,000 mile service.

5 Ignition timing - static

1 The ignition timing must be correct to within about 2°. Variation between different contact breakers, and on one between new and old, alter the timing by more than this. The timing should therefore be checked whenever the contact breaker is adjusted, cleaned, or renewed, and so anyway becomes part of the 6,000 mile maintenance task.

2 The timing is normally a simple adjustment. However, if the complete distributor drive has been removed, refer to section 6 for the initial setting on reassembly. Normal readjustment is done as follows:

3 Having set the contact breaker, turn the engine on forwards, clockwise, till the timing notch in the pulley rim is uppermost. Measure 13 mm circumferentially around the rim, clockwise, and make a mark in the rim: It will be worth filing another notch, but do check it is in the correct place (photo).

4 Now turn the engine on forwards again till this new mark, which shows 10° BTDC, is lined up with the arrow on the timing chain cover. Always turn the engine clockwise, never backwards. If you overshoot, go on another turn, otherwise backlash in the distributor will affect the timing, (photo).

5 The engine is now set at 10° BTDC, so the distributor must be unclamped and adjusted so that the points are just opening. On the sedan the clamp is a nut on the stud on the bottom of

4.3. Set the contact breaker with the heel of the points on the top of the cam

5.3. With the sinple timing mark it is necessary to mark off 13 mm round the pulley

5.4. This FIAT gauge shows what needs to be done

the body of the distributor itself, fixing it to its pedestal sticking out of the crankcase. On the station wagon the stud is on the top of the crankcase.

6 The best way to see when the points are opening is to make a test lamp from a bulb, holder, and two bits of wire, and put this in the wiring from coil to contact breaker. If not doing this, then watch the points very carefully. Turn the distributor body gradually to the point where the points just open. Then tighten the distributor holding nut, without disturbing it.

7 To recheck, now turn the engine over, nearly one complete turn. Ignore the timing mark, go on till the points just open again. Now see where the timing mark is. The first time you may be out, as moving the distributor to and fro might have got the backlash wrong. If necessary, hold the distributor spindle in the retarded position.

6 Distributor - dismantling and refitting

1 Unclip the distributor cap and push it to one side.
2 Undo the low tension wire from the terminal on the side.
3 Undo the nut on the downward facing stud, just below the lubricator. It is in the slot giving the timing adjustment. Remove this nut, and washers, and draw the distributor body up out from the drive pedestal on the crankcase. Note that on the sedan the drive is by a slot in the lower shaft engaging with two tongues in the actual distributor shaft. This is assymetrical, so that the distributor cannot be refitted the wrong way round, (photos).
4 On the station wagon this is different, see paragraph 11 onwards,
5 Remove the fixing screw for the condenser. Take the nut off the terminal bolt, and take off the lead for the condenser and the insulating washers, and the bolt and paper gasket from inside the distributor body. Note the way the large insulating washer on the outside is fitted, with its step ledge through the hole in the casing, to hold the bolt central and away from metal to metal contact. The paper gasket goes between the spring blade and the distributor body, (photos).
6 The contact breaker plate can now be lifted clear after undoing the two screws that go through the spring clips for the distributor cap, (photos).
7 To remove the centrifugal advance mechanism, prise out with a screwdriver the felt pad in the top of the spindle, to uncover the screw. Hold the bottom of the distributor drive shaft in a soft jawed vice, and unscrew the screw at the top. Note the spring underneath.
8 Unhook the automatic advance springs from the two pegs at the bottom end of the shaft with the contact breaker tag. Now the cam can be slid upwards off the spindle, and the bob weights removed as well. Note carefully the position of each.
9 Reassembly of the sedan is the reverse. The timing will only have been slightly disturbed, so can be reset as described in section 5.
10 However if on the sedan the pedestal with bottom half of the distributor drive shaft, has been removed, it is all too easy to reassemble engaging the teeth on the distributor with the wrong ones on the camshaft. The same situation occurs whenever the station wagon's distributor is removed, as it has a one piece shaft.

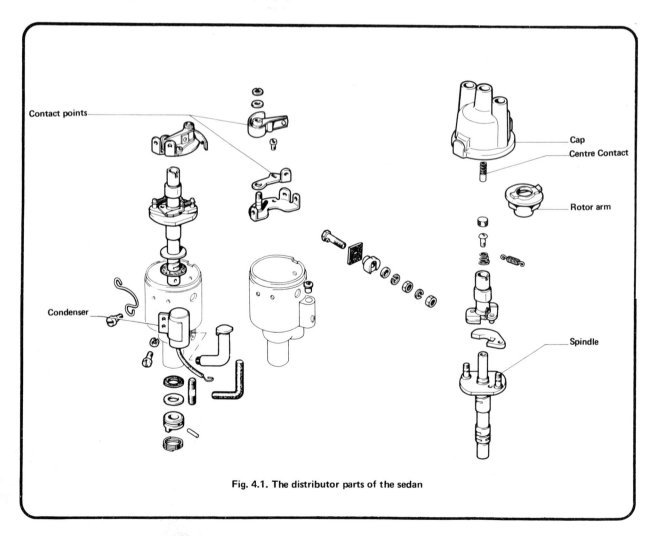

Fig. 4.1. The distributor parts of the sedan

6.3a. If you want to avoid full retiming don't take out the complete distributor shaft unless you need to 'X' marks engine attachment lug

6.3b. Rather, take off the top only, leaving the gear wheel and lower shaft in the crankcase. (Sedan only)

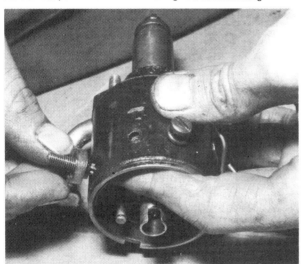

6.5a. Take the nuts and insulator off the terminal bolt

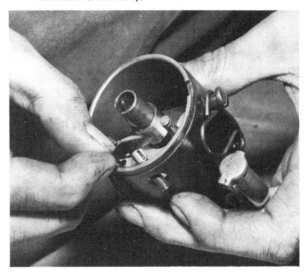

6.5b. Take out the bolt and sheet of insulating paper

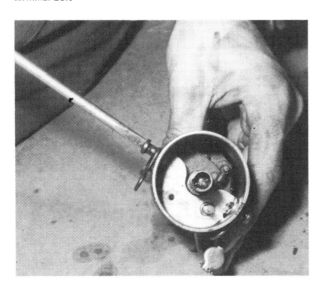

6.6a. Undo the two screws

6.6b. and take out the contact breaker plate

11 To engage the teeth correctly proceed as follows: Turn the engine over till the pulley timing notch is in line with the arrow on the timing chain cover. The engine is now at TDC.
12 Take off the rocker box. Feel the tappet clearances by rocking the rockers. One cylinder will have both rockers free; on the other there will be none, as the exhaust will be just shutting, and the inlet opening. The first mentioned cylinder is ready to fire.
13 Trace back the plug from the cylinder ready to fire to the distributor cap. Note which segment it goes to, and the relationship of that segment to the notch that locates the cap on the body. Fit the rotor arm to the spindle, and turn it till it is lined up for contacting the segment.
14 Now insert the complete distributor drive into the crankcase. The spindle may need a slight turn to engage the teeth. Ensure the fixing for the distributor is lined up with its opposite number on the crankcase as the teeth engage. Now push the distributor home. As you do so the spindle will turn about 30° because of the teeth. Note the angle turned.
15 Pull the distributor out again. Turn the spindle a suitable amount to allow for the twist of engaging the gears. Then push it in again.
16 The distributor should now be in place, the rotor arm pointing to the correct segment in the cap, and the points near enough to opening to be adjustable within the scope of the slot in the upper fixing, as described in section 5.
17 On the station wagon the distributor shaft also has to fit over the splines on the top end of the oil pump shaft. This may involve some extra shifting of the distributor. If necessary turn the engine over a very small amount. Clearly, of course, do not turn the engine over with the distributor out. Put it at TDC before ever taking out the distributor.

7 Ignition advance - testing stroboscopic timing

1 If the working of the automatic advance is suspect it can only be tested by use of a stroboscope, and the FIAT timing plate.
2 A practical way to "check" it is to fit new springs, clean and oil the centrifugal mechanism; then the likelihood of any fault is remote!
3 For those who own a stroboscope the timing is given in the form of a graph in Fig. 4.3.

8 Condenser - removal and replacement

1 The purpose of the condenser, (sometimes known as a capacitor), is to ensure that when the contact breaker points open there is no sparking across them which would prevent proper electromagnetic build up of HT, and cause wear of the points.
2 The condenser is fitted in parallel with the contact breaker points. If it develops a short circuit, it will cause ignition failure as the points will be prevented from interrupting the low tension circuit.
3 If the engine becomes very difficult to start or begins to miss after several miles running and the breaker points show signs of excessive burning then the condition of the condenser must be suspect. A further test can be made by separating the points by hand with the ignition switched on. If this is accompanied by a larger flash than usual it is indicative that the condenser has failed with an open circuit.
4 Without special test equipment the only sure way to diagnose condenser trouble is to replace a suspected unit with a new one and note if there is any improvement.
5 To remove the condenser from the distributor, remove the distributor cap and the rotor arm. Unscrew the contact breaker arm terminal nut, and release the condenser lead. Undo the screw holding the condenser to the outside of the distributor.

9 General distributor wear

1 If the bushes in the distributor are badly worn, the spindle can slop about, so varying the contact breaker gap, and indirectly the timing.
2 New bushes are not available, therefore, a new distributor must be the answer. Note that you do not get one with a reconditioned engine.

10 Spark plugs

1 In the past the plugs were a weak point on many cars, though collecting a still worse reputation because of some very fussy supercharged racing cars. Nowadays they are reliable and long lasting, and in 1972 the normal cost was still the same as in 1938.
2 The plugs should be cleaned at 6,000 miles, and replaced at 12,000 miles. A car burning oil badly, may need more frequent cleaning. One driven hard burns its plugs clean quite well.
3 Cleaning should be done by a garage with a sand blasting machine. Only in the case of the bad oil burning car can satisfactory cleaning be done by hand.
4 The gaps must be set as given in the specification. When setting them never touch the central electrode. Merely bend the outer one.
5 On well used plugs the electrodes burn away, and to a rounded shape, so do not function so well.
6 Only use plugs of the reputable makes and the recommended type. A plug of the wrong type, too cold may foul up. If the other way, too hot, they will not last well, and may even damage the engine.
7 Misused plugs, badly made, or damaged ones, could crack, and drop bits onto the pistons with disastrous results.

11 The coil

1 The only maintenance on the coil is to clean it, and make sure the LT and HT leads make good, secure contact.
2 The fault-finding in Chapter 1 shows how the coil is tested.
3 They only fail occasionally. The most usual reason for failure is leaving the ignition switched on by resistance when the engine is not running. The insulation then breaks down due to overheating. Instead of the LT being cut over half the time by the contact breaker it flows continuously.

12 Fault finding

1 Ignition faults normally result in the engine not running at all; and as a cause of engine failure the ignition system is the most common reason.
2 The tracing of ignition faults stopping the engine were given in the fault finding at the end of Chapter 1. However other faults can appear in the system.
3 Missing.
  If the engine is loosing much power, and sounding very rough, it could be running on only one cylinder. This is usually caused by a faulty sparking plug. To trace which is at fault, wearing gloves, and with the engine running, pull off, and then replace, each plug lead in turn. The plug that makes no difference is not working. The plug should be changed. If this makes no difference check the leads and distributor cap.
4 Wrong timing.
  If the timing is not correctly set many symptoms will appear. If only a few degrees out there will just be loss of acceleration and excessive fuel consumption. If worse, the engine is likely to run hot. If the error is advanced ignition timing the engine will sound rough. Extended running like this can damage the engine.
5 Damp and dirt.
  The most common cause of difficult starting is the combination of dirt and damp on ignition leads, and inside and outside the distributor, and the coil contacts. Aerosol sprays that repel damp allow a start that time, but do not remove the root cause, the dirt.

White deposits and damaged porcelain insulation indicating overheating

Broken porcelain insulation due to bent central electrode

Electrodes burnt away due to wrong heat value or chronic pre-ignition (pinking)

Excessive black deposits caused by over-rich mixture or wrong heat value

Mild white deposits and electrode burnt indicating too weak a fuel mixture

Plug in sound condition with light greyish brown deposits

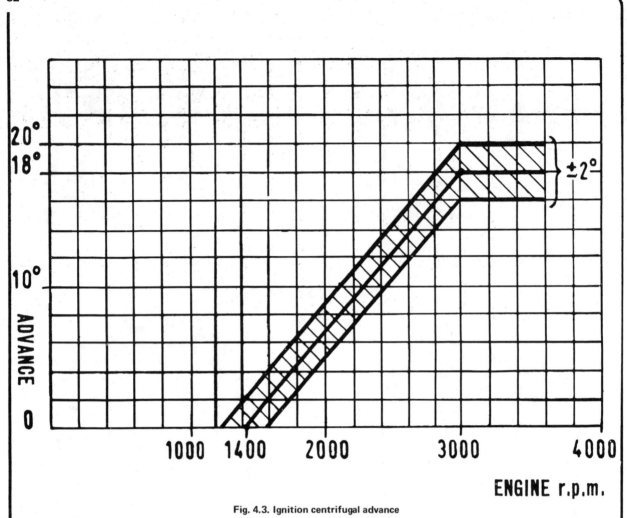

Fig. 4.3. Ignition centrifugal advance

Chapter 5 Clutch

Contents

General description ... 1	Clutch spigot bearing ... 4
Clutch assembly - removal, inspection and replacement ... 2	Clutch cable replacement ... 5
Clutch withdrawal operating mechanism - inspection and repair ... 3	Clutch life ... 6
	Fault diagnosis ... 7

Specifications

Coil spring type up to Sedan 824000 and Station wagon 141256

Type ...	Single-plate, dry disc
Linings, outside diameter ...	(140 mm)
Linings, inside diameter ...	(96 mm)

Clutch springs

Number of working coils ...	5.75
Total number of coils ...	7.25
Free length ...	(40.5 mm)
Seated length ...	(24.5 mm)
Corresponding load ...	(26 ± 1.2 kg)
Minimum permissible load ...	(22 kg)

Release lever carrier ring springs

Number of working coils ...	10½
Seated length ...	(30 mm)
Corresponding load ...	(2.2 ± 0.2 kg)

Clutch pedal free play ...	1.3/8 inch to 1.9/16 inch (35 to 40 mm)
Run-out of driven plate linings, max ...	.0059 inch to .0118 inch (0.15 to 0.30 mm)
Clearance between clutch shaft and driven plate hub splines:	
—endwise ...	.0019 inch to .0039 inch (0.05 to 0.10 mm)
—crosswise ...	.0059 inch to .0118 inch (0.15 to 0.30 mm)

Diaphragm spring type from Sedan 824001 and Station wagon 141257

Throw-out mechanism ...	The diaphragm spring
Linings: outside diameter ...	155 mm
inside diameter ...	114 mm
Clutch pedal free play ...	5/8 inch - ¾ inch (15 - 20 mm)

Tightening torque

Clutch cover to flywheel ...	6 - 7 lbf ft (0.8 - 1.0 kg.m)

Chapter 5/Clutch

1 General description

1 The clutch is a single disc design and incorporates a driven plate (which carries the friction material on each side) and a pressure plate and cover assembly. On early cars the pressure plate is tensioned by coil springs and the pressure is taken off by a central release ring linked to three release levers which pivot on the cover.

2 The clutch operating lever pivots in the forward end of the gearbox casing and a thrust bearing on the inner end bears on to the release ring when the arm is operated. The operating arm is moved by a cable from the clutch pedal.

3 As the friction surfaces of the driven plate wear so the clearance between the thrust bearings and release ring decreases. This clearance is reflected in the free play movement of the clutch pedal. This movement can be adjusted by altering the length of the cable. This is done by turning the adjuster nut fitted to the clutch end of the cable.

4 From the 110F onwards (see chassis numbers in the specification) the clutch was redesigned, and now has one diaphragm spring. The advantage of this is that as the pressure plate moves nearer the flywheel as the friction linings wear the spring tension stays much the same. Pedal pressure releasing the clutch is less. There are fewer parts, and the clutch is easier to balance.

5 The withdrawal bearing is a carbon ring on the coil spring clutches, and a ball bearing on the diaphragm ones. The station wagon withdrawal mechanism is slightly different from the sedans. Clutchares are not interchangeable.

2 Clutch assembly - removal, inspection and replacement

1 Remove the engine as described in Chapter 1.

2 Working diagonally slacken the six mounting bolts which hold the clutch cover to the flywheel - slackening each one a little at a time until the tension on the pressure springs is completely relieved. Lift off the cover and clutch driven plate.

3 The clutch driven plate should be inspected for wear and for contamination by oil. Wear is gauged by the depth of the rivet heads below the surface of the friction material. If this is less than 0.025 inch (0.6 mm) the linings are worn enough to justify renewal.

4 Examine the friction faces of the flywheel and clutch pressure plate for signs of scoring or overheating. These should be bright and smooth. If the linings have worn too much it is possible that the metal surfaces may have been scored by the rivet heads. Dust and grit can have the same effect. If the scoring is very severe it could mean that even with a new clutch driven plate, slip and juddering may recur.

5 Deep scoring on the flywheel face is serious because the flywheel will have to be removed and machined by a specialist, or renewed. This can be costly. The same applies to the pressure plate in the cover although this is a less costly affair. If the friction linings seem unworn yet are blackened and shiny then the cause is almost certainly due to oil. Such a condition also requires renewal of the plate. The source of oil must be traced also. It will be due to a leaking seal on the transmission input shaft (Chapter 6 gives details of renewal) or on the front of the engine crankshaft (see Chapter 1 for details of renewal).

6 If the reason for removal of the clutch has been because of slip and the slip has been allowed to go on for any length of time it is possible that the heat generated will have adversely affected the pressure springs in the cover. Some or all may have been affected with the result that the pressure is now uneven and/or insufficient to prevent slip, even with a new friction plate. It is recommended that under such circumstances a new assembly is fitted.

7 Although it is possible to dismantle the cover assembly of the old coil spring type clutch to renew the various springs and levers, the economics do not justify it. Clutch cover assemblies are available on an exchange basis. The component parts for their overhaul are not readily available, and it is difficult to compress the springs to release the pressure plate from the cover. It is impractical to overhaul the diaphragm spring clutches as they are rivetted together.

8 Clutches are available relatively cheaply. Remembering the great labour in dismantling the engine to get at the clutch it is well worth fitting a reconditioned assembly, new driven plate, and new withdrawal bearing.

9 When replacing the clutch, support the driven plate on a finger through the centre of the withdrawal mechanism. Be sure that the short length of the hub of the disc is towards the flywheel (see section drawings).

10 Stand the six securing bolts into the threads in the flywheel, ensuring that the holes in the clutch cover are lined up with the dowels on the flywheel. Tighten them enough for the linings of the driven plate to be lightly gripped, but still moveable.

11 It is important to line up the central splined hub with the bearing in the bore of the crankshaft. If this is not done it will be impossible to refit the engine to the transmission. It is possible to centralise them by eye but a simple surer way is to select a suitable piece of bar or wooden dowel which will fit snugly into the crankshaft spigot bearing and round which some adhesive tape can be wound to equal the diameter inside the driven plate boss. By inserting this the plate can be moved and centralised with sufficnent accuracy.

12 Finally tighten up the six cover securing bolts evenly and diagonally a little each at a time to a final torque of 6 - 7 lbf ft.

13 Before refitting the engine after a clutch overhaul check the transmission input shaft oil seal (Chapter 6) and the clutch release operating mechanism (see Section 5).

14 Do not put any oil or grease on the splines of the shaft or the clutch driven plate. This would become sticky, and cause clutch drag. But put some grease on the end of the shaft where it goes in the spigot bearing in the crankshaft.

3 Clutch withdrawal operating mechanism - inspection and repair

1 Clutch operation can be adversely affected if the release thrust bearing and retaining springs are worn or damaged. Squeals, juddering, slipping or snatching could be caused partly or even wholly by this mechanism.

2 Full examination is possible only when the engine has been removed and normally it is carried out when the clutch is in need of repair. The mechanism is contained in and attached to the transmission casing.

3 On coil spring type clutches the withdrawal bearing is a carbon ring. They do not last very long. If there is any doubt about it, fit a replacement. They are simply held into the withdrawal fork.

4 On diaphragm clutches the withdrawal bearing is a ball race. Unless you remove the complete assembly from the transmission housing and strip it do not wash it lest grease be washed out without the possibility of replacing it.

This should only need doing when the transmission is being given a complete overhaul.

5 Check that the actuating mechanism moves freely.

6 Readjust the free play when the engine is reinstalled.

7 During reinstallation take great care not to damage the clutch with the gearbox input shaft whilst the engine is being fitted to the transmission.

4 Clutch spigot bearing

1 The engine-end of the clutch shaft's rests in a spigot bearing in the crankshaft. See Fig. 1.1 and 1.2.

2 After very lengthy service it wears. This allows the shaft to move off centre. The oil seal in the flywheel housing where the shaft goes into the transmission will not last long with this running out of true. It could also result in rough running due to out of balance.

3 To check the bearing the shaft must be removed from the transmission, and the clutch from the engine. It can therefore

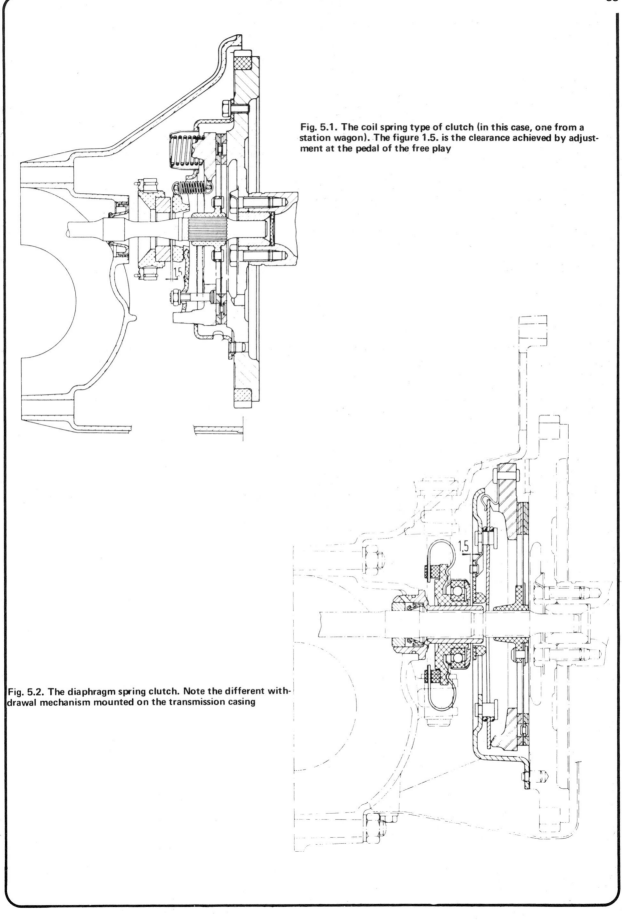

Fig. 5.1. The coil spring type of clutch (in this case, one from a station wagon). The figure 1.5. is the clearance achieved by adjustment at the pedal of the free play

Fig. 5.2. The diaphragm spring clutch. Note the different withdrawal mechanism mounted on the transmission casing

only be done when the transmission is being dismantled.
4 The bearing is a push fit in the crankshaft. The flywheel must be off to replace it.
5 To extract the old one, ideally tap a screw thread, or use an "Easi-out" and pull it out. We pulled ours when the crankshaft was out by putting the end of the bearing in the vice.
6 Lubricate the new one, and tap it into place.
7 The recommendation is to replace the spigot bearing if the crankshaft is out of the engine, as if the engine is worn, the spigot probably is too. The spigot bearing is cheap, so it is worth taking the opportunity to change it when it occurs.

5 Clutch cable replacement

1 The cable should only need replacement if it breaks. Stretch should not be severe.
2 However, should the cable appear too long, perhaps after fitting a new clutch, and the threaded part on the end is reached whilst the free play is too large, then put either a short length of tube, or large nuts, over the adjuster rod between the clutch lever and adjuster nut.
3 To change the cable, first jack up the car and support it on blocks.
4 Remove the cover over the spine of the car on the floor behind the front seats.
5 Remove the heater trunking on the floor spine beside the drivers pedals.
6 At the rear unhook the pull off spring from the clutch lever, and unscrew the lock and adjuster nuts from the cable ends.
7 At the front remove the split pin and washers holding the cable end to the pedal.
8 Withdraw the cable underneath the car. The outer cable will have to be disengaged from the seats in the bulkheads at each end of the spine in the floor of the car, and slid sideways along the slots cut to the larger holes.
9 The new cable is supplied complete with rubber dirt excluder and seats.
10 Readjust the free play at the pedal.

6 Clutch life

1 Some drivers allow their clutches to outlast the engine and transmission. Others need new ones almost as part of the 12,000 mile service. The clutch is one of the prime examples of a component cheap to buy, but expensive (or lengthy for the home mechanic) to fit.
2 The short life is due to abuse.
3 Common abuses can be avoided as follows:
4 Do not sit for long periods, such as at traffic lights, with the clutch disengaged. Put the gearbox in neutral until the lights go yellow. This extends withdrawal bearing life.
5 Always remove the foot completely from the clutch pedal once under way. Riding the clutch lightly ruins the withdrawal bearing and thereby the rest of the clutch.
6 Whenever the clutch is disengaged, hold it completely so, the pedal down as far as it will go. You see drivers holding the car on a hill by slipping the clutch. This wears it out quickly.
7 Always move off in 1st gear.

Fig. 5.3. The coil spring clutch installed on the engine

Fig. 5.4. The clutch assembly removed, with driven plate beside it, both as seen from the engine

Fig.5.5 The diaphragm spring type of clutch during assembly
1 Clutch pressure plate
2 Clutch centralising mandrel
3 Flywheel

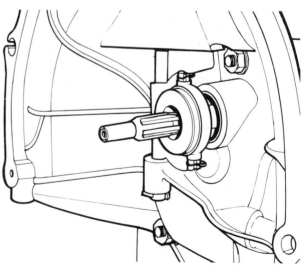

Fig. 5.6. The clutch withdrawal mechanism

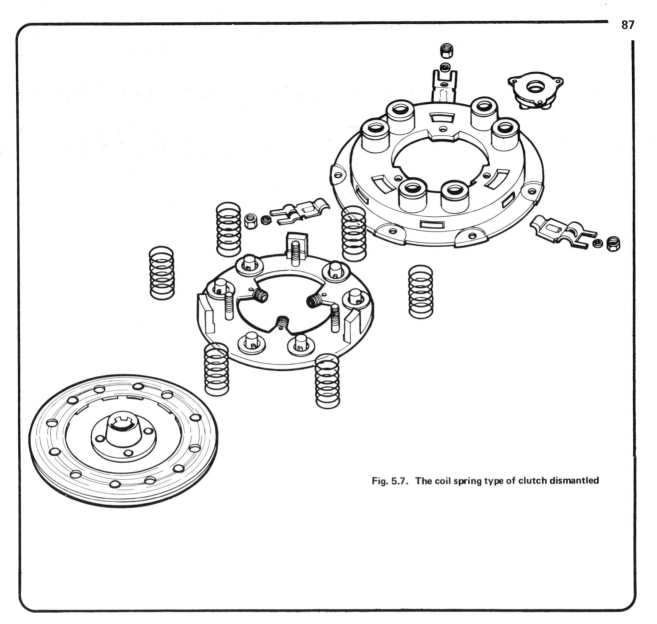

Fig. 5.7. The coil spring type of clutch dismantled

Fig. 5.8. Spanners on adjuster nut and locknut of the clutch cable

Fig. 5.9. Clutch pedal and cable fixing

7 Fault diagnosis and remedies

Symptom	Reason/s	Remedy
Judder when taking up drive	Loose engine/gearbox mountings or over-flexible mountings	Check and tighten all mounting bolts and replace any 'soft' or broken mountings.
	Badly worn friction surfaces or friction plate contaminated with oil carbon deposit	Remove engine and replace clutch parts as required. Rectify the oil leak which caused contamination.
	Worn splines in the friction plate hub or on the gearbox input shaft	Renew friction plate and/or input shaft.
	Badly worn bearing in flywheel centre for input shaft spigot	Renew bearing in flywheel.
Clutch drag (or failure to disengage) so that gears cannot be meshed	Clutch actuating cable clearances too great	Adjust clearance.
	Clutch friction disc sticking because of rust on splines (usually apparent after standing idle for some length of time)	As temporary remedy engage top gear, apply handbrake, depress clutch and start engine. (If very badly stuck engine will not turn). When running rev up engine and slip clutch until disengagement is normally possible. Renew friction plate at earliest opportunity.
	Damaged or misaligned pressure plate assembly	Replace pressure plate assembly.
Clutch slip - (increase in engine speed does not result in increase in car speed - especially on hills)	Clutch actuating cable clearance from fork too small resulting in partially disengaged clutch at all times	Adjust clearance.
	Clutch friction surfaces worn out (beyond further adjustment of operating cable) or clutch surfaces oil soaked	Replace friction plate and remedy source of oil leakage.
	Damaged clutch spring(s)	Fit reconditioned assembly.

Chapter 6 Transmission

Contents

General description ... 1	Meshing the crown wheel and pinion ... 8
Removal of the transmission from the car (engine in place) .. 2	Reassembling the gearbox ... 9
Removal of drive shafts and differential ... 3	Reassembly of the differential and drive shafts ... 10
Stripping the differential ... 4	Refitting the transmission to the car ... 11
Stripping the gearbox ... 5	Gear lever adjustment ... 12
Renovation of the gearbox ... 6	Smooth silent gear changing ... 13
Renovation of the final drive ... 7	Fault finding ... 14

Specifications

Ratios:	New 500	500 Sport	All later cars
1st gear	3.27	3.27	3.70
2nd gear	2.06	2.06	2.06
3rd gear	1.30	1.30	1.30
4th gear	0.87	0.87	0.87
Reverse	4.13	4.13	5.14
Final drive	5.125	4.88	5.125
1st gear overall	16.8	15.95	18.96
2nd gear overall	10.6	10.1	10.6
3rd gear overall	6.7	6.3	6.7
4th gear overall	4.5	4.3	4.5
Reverse	21.2	20.2	26.33

Oil capacity ... 1 qrt (1.26 US qrt) 1.1 litre

Gearbox
Condemnation limits

Input shaft bearing:	Selector-end: Sideways play	.0018 inch (.045 mm)
	End float	.0177 inch (.450 mm)
	Differential-end: Sideways play	.0016 inch (.040 mm)
	End float	.0156 inch (.400 mm)
Output shaft:	Selector-end: Sideways play	.0016 inch (.040 mm)
	Differential-end: Sideways play	.0018 inch (.045 mm)

Backlash between gears : maximum ... .004 inch (.10 mm)
Gear wear limit on bush ... .008 inch (.20 mm)
Dog-clutch sliding sleeves on hubs ... .006 inch (.15 mm)

Final drive
Differential side bevel gear thrust ring:

	Standard	.039 inch (1 mm)
	Oversizes various up to	.051 inch (1.3 mm)

Crownwheel and pinion distance "B"
Which is differential centre line to seat of pinion shoulder on inner
race of output shaft bearing ... 75 mm
Shims available in thicknesses: ... .10 and .15 mm
Crownwheel to pinion backlash ... .08 - .12 mm (.0031 inch - .0047 inch)
Differential bearing preload: Torque to rotate ... .130 - .150 kgm (.94 - 1.08 lbf.ft)

Shaft wear limits:
Drive shaft cross head slip joint in differential
bevel side gears ... ... .20 mm (.0079 inch)
Shaft in sleeve at hub end ... ... .15 mm (.0059 inch)

Tightening torques
Input shaft end nut ... ... 18½ - 25 lbf.ft (2.5 - 3.5 kg.m)
Output shaft end nut ... ... 29 - 36 lbf.ft (4 - 5 kg.m)
Bolts through crownwheel in differential casing ... ... 23½ lbf.ft (3.2 kg.m)
Flywheel housing to gearbox casing ... ... 27½ lbf.ft (3.8 kg.m)
Transmission to engine nuts ... ... 18 - 22 lbf.ft (2.5 - 3.0 kg.m)
Drive shaft sleeve to hub ... ... 20 lbf.ft (2.8 kg.m)

1 General description

1 The gearbox and final drive with differential are built as one transmission unit. The drive from the clutch passes forward by a long shaft over the differential into the gearbox. On this shaft in the top of the gearbox is a gear cluster with a series of gearwheels, one for each gear, and all permanently locked to their shaft, the input shaft. Below is the output shaft. On this lower shaft again there is a gearwheel for each gear speed, but these being in constant mesh with those on the top, input shaft, are normally free, and just one locked to the output shaft when needed. The output shaft brings the drive back again to the final drive, where it is turned across the car and given another gear reduction.

2 The gears are engaged by locking them to the output shaft with dog-clutch sleeves sliding on splined hubs on the output shaft. There is not any synchromesh to get the gearwheel going at the correct speed before the dog-clutch engages. This makes work on the gearbox much simpler. The gearwheels are very light on such a small gearbox, so if gear changes are made without much skill the teeth of the dog-clutches do not complain loudly, but changes should be made employing double-declutching.

3 The price of a reconditioned transmission at first looks rather high but for this you get a guaranteed factory overhaul. If the transmission has one isolated and obvious defect, like running very rough in one gear indicating the gear teeth have failed, it could readily be stripped and that one defect rectified. The problem comes if work is undertaken involving changing the crownwheel and pinion, the double ball thrust bearing at the front, or selector-end, of the output shaft. Unless the official workshop setting gauges are available it is difficult to set up the position of the pinion, which is also the output shaft, or mesh the crownwheel properly, and adjust the preload of its bearings. If just general noise is the problem a partial overhaul may give disappointed results. Yet if every component is to be renewed then the bill will be expensive, and there could well be difficulty in getting all the parts. But if silence is not the problem, then everything is fairly straightforward.

4 When you strip the transmission it is important to lay out the components in order as they come out or off, so that they can go back in the same order, and in the same relative position, and not get muddled up.

2 Removal of the transmission from the car (engine in place)

1 It is normally expected that if the transmission is needing repair, some other components of the car will be as well, and therefore the transmission will be removed with the engine. This makes work a lot easier, as there is less to do under the car, and the car has not got to be lifted to a height to clear the transmission as it comes out (if there is no pit for working under the car). The removal with the engine is described in Chapter 1.

2 If it is necessary to remove the transmission with the engine still in place, proceed as follows.

3 Place the car over the pit, or jack it up well clear of the ground, and mount it very securely, with the wheels clear, so that they can be turned over.

4 Drain the oil from the transmission.

5 Remove the three bolts securing each drive shaft sleeve to the hubs. Reach inside the joint and remove the small spring between the hub and the drive shaft.

6 Disconnect the starter motor, and all controls on the transmission and the speedometer cable, as described in Chapter 1.5.

7 Remove the starter motor.

8 Support the engine by a jack, putting wooden blocks between the jack head and the sump. Undo the ring of bolts securing the flywheel housing to the crankcase. Put a second jack under the transmission.

9 Undo the two bolts holding the front crossmember and transmission mounting to the car body.

10 Lower the jacks very slightly, and then slide the transmission forward so the shaft comes out of the clutch driven plate. Then lower the transmission clear of the car. In doing this there might be some trouble in getting the drive shafts clear of the axle. If you are working alone these might need wedging clear of the suspension arms whilst the transmission is manoeuvred out.

11 Clean the exterior of the transmission.

12 Remove the transmission mounting support bracket and rubber mountings from the casing.

Fig. 6.1. The transmission showing the gearbox and final drive parts

1 Pin
2 Shaft joining sleeve
3 Joining sleeve lock ring
4 Clutch shaft seal
5 Clutch shaft
6 Rear roller bearing
7 Shims
8 4th speed driven gear
9 1st speed and reverse sliding gear
10 3rd/4th speed engagement sliding sleeve
11 Hub for sleeve
12 3rd speed driven gear
13 Front ball bearing
14 2nd speed driven gear
15 2nd speed engagement sliding sleeve
16 Speedometer drive gear
17 Output shaft
18 Speed selector lever
19 2nd speed drive gear
20 Front ball bearing
21 Input shaft
22 Rear ball bearing

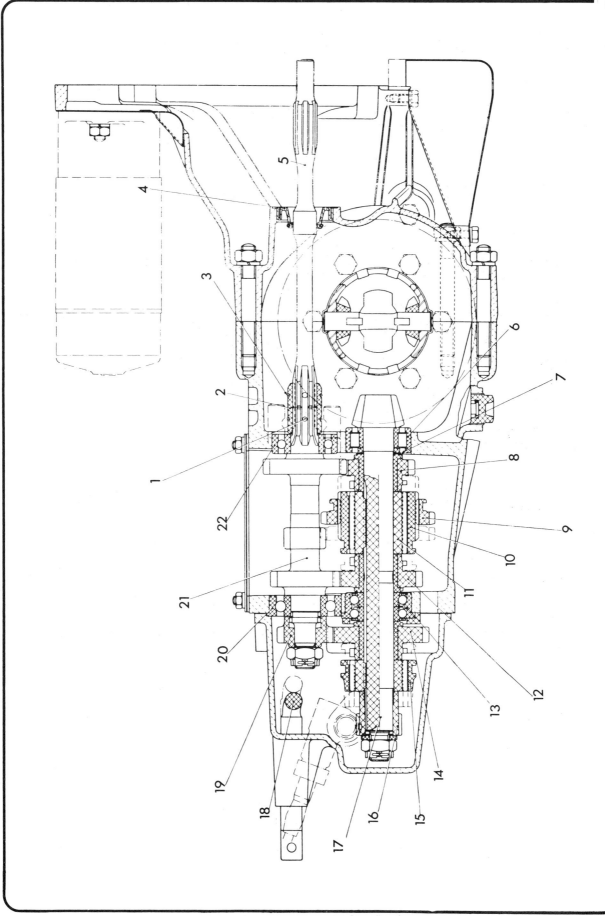

91

3 Removal of drive shafts and differential

1 To remove these the transmission must be removed from the car and the differential dismantled.
2 Remove the driving sleeves from the outer ends of the drive shaft. To do this prise back the rubber dirt excluder, and push that to the middle of the shaft. Clean any dirt that might have got under the excluder onto the splines, and then push the drive sleeve along the shaft so that the circlip at the end is exposed. Remove the circlip, and then slide the sleeve off, with its dirt excluder, (photos).
3 Remove the four bolts holding the ring on the rubber oil retaining boots where the driveshafts fit into the transmission casing, (photo).
4 Take off the boots noting that inside is a locking tab that holds the adjustment for the output bearing race, (photo).
5 Undo the four nuts holding the star shaped outer race housing for the output bearing to the transmission casing, and take off the race housing, (photo).
6 Repeat for the other side.
7 Mark the position in which the race housings were fitted on the studs, so that they are replaced in the same way.
8 Note that in the housing is the castellated ring that sets the bearing preload for the differential. This adjusts the meshing and preload of the differential, and it is important that it is not disturbed. Mark the position of the castellated ring in the bearing housing with paint if it appears free to move.
9 As the star shaped bearing housings have been removed, the differential section of the housing can now be split.
10 Remove the nuts from the studs in the flywheel housing either side of the clutch withdrawal mechanism, which hold the two halves of the differential section of the transmission casing together, (photo).
11 Remove the flywheel housing from the rest of the transmission casing, (photo).
12 Holding the two drive shafts lift the differential clear of the casing, (photo).
13 Now strip the differential as described in section 4 following, when at last the drive shafts will be free.

4 Stripping the differential

1 In order to remove the drive shafts from the differential, the latter must be dismantled.
2 Undo the bolts holding the crownwheel to the differential case. Hold the differential whilst this is being done either in a vice with soft jaws, or by passing a bar through the hole in the casing.
3 The crownwheel is now free from the casing, and the two halves of the differential casing can be split. Mark every item as you take it apart so that you can refit it in the same position as it was before. The two halves of the differential casing are likely to be a tight fit and will require careful prising apart to keep them straight as they are withdrawn, (photo).
4 The drive shaft with the driving cross heads of the slip joint can now be taken apart, (photo).
5 Note the bronze thrust ring that takes the outward thrust of the differential bevel side gears against the end of the casing. These are of varying thickness to give correct meshing of the bevel gears, (photo).
6 The right hand bevel side gear and drive shaft cannot come out until the two idler pinions on their shaft are removed. To release these bend back the lugs holding the shaft retaining cup to the differential casing, and prise it off carefully, so that it can be reused.
7 Leave the roller bearings in place on each half of the differential casing, and their outer races in the housings unless they are to be renewed.

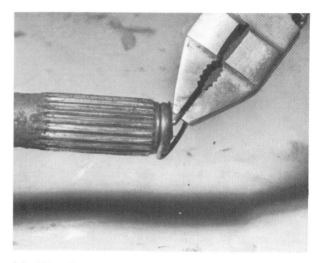

3.2a. Take off the circlip,

3.2b. the coupling sleeve,

3.2c. and the dirt excluder from the outer ends of the driveshafts

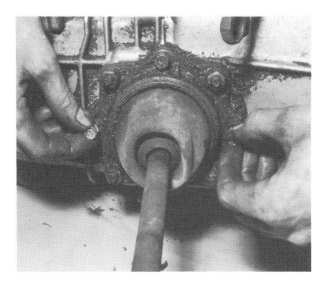

3.3. Take off the bolts for the oil-boot retainer,

3.4. which also holds the locking-tab ring

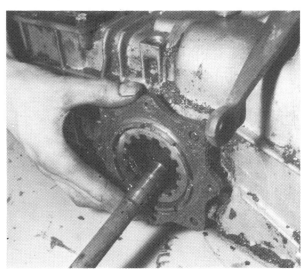

3.5. Carefully prise off the bearing housing

3.10. Remove the six nuts and washers,

3.11. and pull the flywheel housing off the gearbox

3.12. Lift out the differential by the drive shafts

4.3. Undo the bolts, then prise the differential apart

4.4. The cross-heads on the drive shaft are now free to slide out of the bevel side gear. The grooves in the working surface assist lubrication

9 Remove the two nuts holding the cover for the springs for the gear selector detents.

10 Remove the three springs from their holes. Tip up the gearbox so that the three detent balls roll out of their passages, (photos).

11 Remove the bolt holding the reverse selector fork to its rod, slide the rod out of the gearbox casing, and lift out the fork, (photo).

12 In the same way remove the bolt holding the 3rd/4th gear selector fork. Pull the rod out a bit, and get from the gearbox casing one interlock ball and one thin plunger from between the housings for the far end of the gear selector rods. Take out that selector rod and its fork, (photos).

13 From the outside of the casing remove the bolt locking the reverse gear shaft, and then from the differential end of the casing push the shaft out towards the selector-end. Lift out the reverse gear pinion, (photos).

14 Slide two gears into mesh at the same time, so that the gearbox is locked.

15 Remove the split pin from the end of the output shaft, and undo the nut. Do the same for the input shaft, (photos).

16 From the lower, the output shaft, remove the washer and then holding a hand underneath it, to catch the ball that is its locating member instead of a key, slide the speedometer drive pinion off the output shaft, (photo).

17 Take the second gear pinion off the end of the input shaft.

18 Undo the bolt holding the first gear selector fork to the lowest gear engaging rod. Slide out the rod, pulling it by the second gear fork at the end, and bringing off the end of the output shaft the second gear engaging sleeve which is within the arms of the fork. Look for and catch the last ball from the interlock mechanism in the differential-end housing for the rod, (photos).

19 Take off the output shaft the remaining parts: The hub for the engaging sleeve, the second gear wheel and its bush. (photo)

20 Undo the two screws countersunk into the selector-end bearing housing for the output shaft. These may need tapping with a punch to start them out of the housing. Take off the bearing housing, (photo).

21 The selector-end bearing for the input shaft is now free. If it will come easily, just take it out: Note the flange with the flat cut on it to clear the other bearing, (photo).

22 Slide the input shaft towards the selector end; out of the end of the casing. It may need carefully driving out with a drift. If the selector-end bearing could not be got out before, it will be driven out now. When the shaft is free pull that selector end bearing off the end of the shaft. It does not matter if the differential-end bearing is still on the end of the input shaft, or left in the casing.

4.5. The thrust rings in the differential case for the side gears also adjust end-float

5 Stripping the gearbox

1 The front half of the casing with the differential having been removed as described in the previous section, proceed as follows:

2 Pinch together the ears of the circlip on the sleeve locking the clutch shaft onto the gearbox input shaft. Slide the circlip out of its groove and let it rest on the sleeve. Now push out the pin that is through the sleeve and through the input shaft. (This is the furthermost of the two pins). Pull the clutch shaft and sleeve off the gearbox input shaft, (photo).

3 Remove the four set screws holding the gearbox top cover, and take off this cover from the gearbox, (photo).

4 Turn the gearbox on end.

5 Remove the nut from the stud holding the speedometer drive, and withdraw the drive, (photo).

6 Undo all the nuts holding the gearbox selector rod end cover to the main casing.

7 Pull the end cover upwards.

8 Lift the end casing off, sliding it along the gear selector rod. If there is difficulty in getting the selector rod to move down into the housing, it may be necessary to file off any burrs on the end of the selector rod shaft. If it gets stuck it is possible to lift off the end cover with the selector rod still in place, unhooking it from the gear shift rods, (photo).

5.2a. Spring aside the far circlip,

5.2b. and take out the far pin, so the clutch shaft will come off bringing the sleeve

5.3. Take off the cover

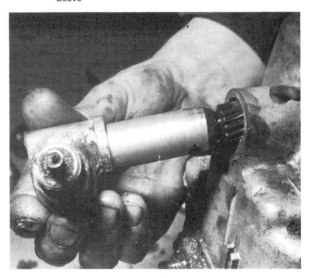

5.5. Remove the speedometer drive

5.8. and the end cover. Our gear selector wouldn't slide out, so we unhooked it from the rod ends

5.9. Take off the detent spring cover

5.10a. Take out the springs

5.10b. Tip out the three balls. Remember the way the selector rods face the detents for reassembly

5.11. Unbolt the reverse fork and pull out the rod

5.12a. With the next rod,

5.12b. comes the interlock plunger,

5.12c. and the first of the two elongated balls

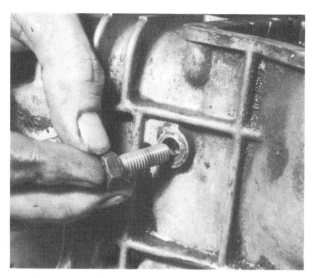

5.13a. The reverse shaft lock-bolt comes out of the casing side

5.13b. Then the shaft can be driven out. It will NOT go the other way

5.13c. The reverse gear is free

5.15a. The nut, without washer, and the 2nd gear pinnion on the input shaft

5.15b. The nut and washer on the output shaft

5.16. Don't loose the ball for the speedometer drive

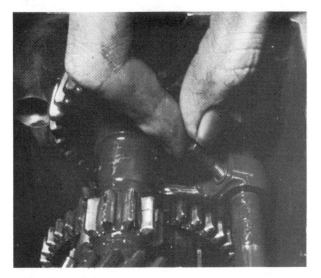

5.18a. Unbolt the 1st gear fork

5.18b. The rod brings off the 2nd gear fork: there will also be the last interlock ball at the other end

5.19. The gear-engaging dog-cluster hub and the 2nd gear wheel

5.20. Take off the output shaft bearing housing

5.21. Then the input shaft bearing: If it is stiff drive it out with the shaft from the other end

5.23. The input shaft being threaded out with the differential-end bearing on it

Chapter 6/Transmission

23 Lift the input shaft out of the casing. If the bearing is on its differential-end, it can just be threaded out through an area in the opening where the side has been machined to allow this. Leave the bearing where it is for now. It need only be moved to replace it if it is worn out, (photo).

24 Now take out the output shaft. Pull it slowly by the pinion out of the differential-end of the casing. As it moves each gearwheel and engaging sleeve is going to drop off the shaft. And last there will be the shims next to the bearing. Take all these parts out and lay them out in order and the same way round so that they will not get muddled. The shims are very important. The differential-end bearing should come out on the shaft. (photos)

25 Take out of the selector-end of the casing the other bearing. But if it proves stubborn leave it there lest it be damaged, and it need only come out if to be replaced, (photo).

26 If the parts cannot be left out in neat order due to lack of space it is suggested they are all put back on the shafts to store them in correct order and the right way round, (photo).

5;24a. Draw out the output shaft

5.24b. with bearing and shims still on it. Or else they will be in the box with the gears

5.24c. The other parts come off inside the casing and can be lifted out

5.25. The selector-end bearing of the output shaft locates the pinion in relation to the crown wheel

5.26. This is what came off the two shafts. It goes back in the reverse sequence

6 Renovation of the gearbox

1 Clean all parts, examining them for signs of damage; scuffing, scoring, cracks or chips out of gear teeth. Check all ball and roller bearings for smooth rotation, and any visible defects: These last must be very carefully washed quite clean in clean paraffin.

2 Put each gear wheel on its own hub and check the pair for wear. Check the ball and roller bearings' free play against the specifications, by judging it, by "feel". If your dealer is co-operative, compare them with new ones.

3 Temporarily reassemble the components that go on the input and output shafts into the gearbox, and check the gear pairs for excessive backlash.

4 If the double ball bearing at the selector end of the output shaft is replaced this is going to upset the position of the pinion in relation to the crown wheel, as it is this bearing that takes the end thrust. See Section 8.

5 Also if the bushes supporting either 3rd or 4th gear wheels on the output shaft are changed there is the same risk. In this case compare the lengths of the old and new bushes by standing them on end on a flat surface, such as a sheet of glass, and putting a steel ruler across the top so that the difference can be measured with feeler gauges. It can be reckoned that each bush will shrink about ½ a thousandth of an inch (.0005 in) when first fitted. So they can afford to be a little larger. If there is greater variation alter the shims at the pinion accordingly.

6 Pay particular attention to the 4th gear wheels as these are used for such a longer distance than the others.

7 You will need new gaskets for the cover, end part of the casing, speedometer drive, and plate retaining the gear detent springs and balls. You will need new oil seals for the input shaft from the clutch, and new boots and dirt excluders for both ends of the drive shafts.

7 Renovation of the final drive

1 Your diagnosis from the noises made when the car was in use should have led you to suspect what major components in the final drive seemed at fault. In particular a loud whine might demand the crown wheel and pinion be changed even though they look all nice and smooth.

2 Examine all parts for signs of damage: Scoring, cracks etc. In particular look at the races and the rollers for the taper roller bearings that support the differential in the casing. These should be smooth and shiny. There is a strong possibility they will appear pitted, almost as if rusting has taken place. If so they will need replacing. They can be a loud source of noise.

3 If there has been loud whining for some time, and the car has done a high mileage, then the crown wheel and pinion will need replacement. But if the mileage is low, and the onset of whine recent, then it could be possible to silence it by resetting the meshing. The two have to be replaced as a pair.

4 If the crown wheel and pinion are not being replaced it is vital that the adjuster rings in the star shaped bearing houses are not disturbed, or the meshing adjustment and bearing preload will be changed.

5 Measure the thickness of the bronze thrust rings in the differential casing for the bevel side gears. If these are worn more than 0.001 inch replace them. Do so with the oversize ones.

6 Check the clearance of the crossheads on the drive shafts in the bevel gears. If slightly excessive new crossheads may suffice. Otherwise the gears must be replaced too.

7 Check the slop of the sleeve on the hub end of the drive shaft. If it is an old car, and the splines have got rusty new shafts may be needed. These will give new splines at the outer ends, and also new pivots at the inner end for the crossheads. All this will remove snatch in the transmission due to all the lost motion adding together.

8 New dirt excluders will be needed for the outer ends of the drive shafts. New oil retaining boots will be necessary at the inner ends, and a new oil seal for the clutch input shaft through the flywheel housing.

8 Meshing the crown wheel and pinion

1 If a new crown wheel and pinion have been fitted their meshing will need setting. The same should be done if the position of the existing pinion has been altered by new gearbox parts that affect this.

2 If the meshing is altered care must be taken not to change the bearing preload. If the differential taper roller bearings are replaced the preload must be reset.

3 FIAT do the meshing using dummy shafts and a special gauge. If possible get a dealer to do the setting for you. Assuming this is not possible, it can be done, with difficulty and risk of ending up inaccurately set, by the method that follows:

4 On the end of the gearbox output shaft, which is also the pinion, will be found two figures. One is a mating code for the crown wheel. The other will have a + or - sign before it. This figure is, for that particular pinion, the variation it must have for correct meshing from the nominal 75 mm distance from differential centre line to shoulder at the back of the pinion.

5 If you are only replacing the crown wheel and pinion, then the new pinion can be positioned by changing the thickness of the shims behind the bearing next to the pinion the same amount as the difference between the two figures stamped on the old and new pinions. Then all that need be done is to move the crown wheel across to get the correct backlash, whilst not upsetting the bearing preload. This will then have properly positioned the new pinion, as the altered thickness of shims will have accounted for the variation in the pinions. It can be done on the basis that the remaining parts in the gearbox will have maintained their position.

6 If some of the parts that govern the position of the pinion have been changed then it ought to be repositioned by reference to the centre line of the differential. Without either FIAT's special gauges, or a surface table and proper measuring instruments this cannot be done by measurement. It could be done from first principles using blue marking dye and shifting the two mating gears about till the correct tooth contact is obtained. This needs a lot of practice and is outside the scope of this book to teach. Therefore, in default of the gauges it is recommended that the new pinion/output shaft is fitted, with the shims corrected as already described for both the different number on the pinion, and for any variation in the 3rd or 4th gear bushes' length. It must be assumed that if a new bearing has been fitted, it is the same dimensions as the old.

7 Assemble the gearbox as described in section 9. But at this stage only do the output shaft, just in case it has to come apart again: Leave out the input shaft, but fit the output shaft bearing housing, and tighten the two countersunk screws. Fit the second gear wheel bush, the hub for its sleeve, and the speedometer drive gear, not bothering for now with its locating ball, or the second gear wheel, as these will have to come off again when the time comes to fit the input shaft. Fit the washer and nut on the end of the shaft.

8 Assemble the differential. This might as well be done now in full, putting in the drive shafts. See section 10.

9 Put the differential in position in the casing, with the taper roller bearings, and bolt on the flywheel cover part of the casing, and fit the nuts and tighten these to the correct torque.

10 Fit the star shaped bearing housings to each side of the differential. If new bearings have been fitted leave the adjusters so that the bearings are slack whilst the housings are tightened. If the old bearings are still being used leave the adjusters, but as they are tightened check that the crown wheel is not being forced down on the pinion, and that there is some backlash between them. If there is no backlash due to differences between the old crown wheel and the new then the bearings must be temporarily shifted to allow their fitting. Screw both serrated rings in the same direction the same number of serrations, so that the same preload is kept on the bearings. As the bearing housings are tightened into place turn the differential over so that its bearings may roll into place.

11 Once all is tightened the sideways position of the crown

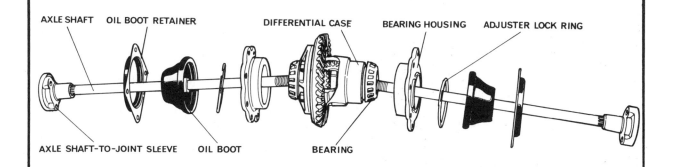

Fig. 6.2. The differential and drive-shaft parts

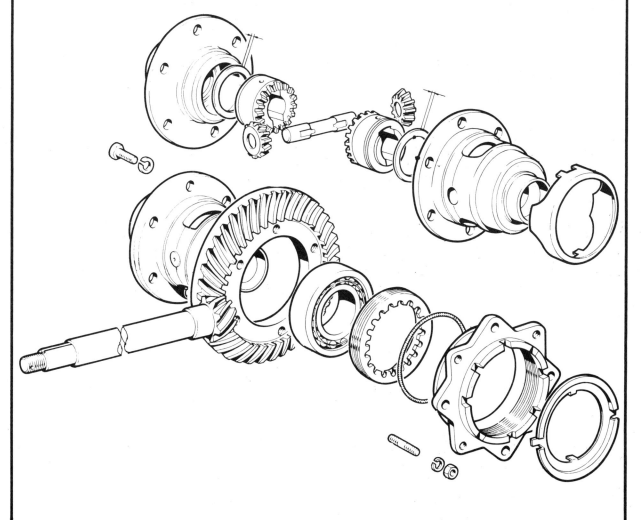

Fig. 6.3. Differential components analysed further

wheel must be adjusted to get the backlash correct. This is difficult to judge as there is access only through the hole for the clutch shaft to the input, admittedly still without the oil seal at this stage.

12 If new bearings have been fitted, or the adjustment for the old ones disturbed, the preload of the bearings must be set.

13 Rotate the differential, by putting the drive sleeves on the splines at the drive shaft ends. Check that there is some backlash between the gears. Continue to turn over the shaft, so that the rollers in the bearings can roll up their races into position, tighten the right sides adjustment serrated ring; this is the one that pushes the crown wheel away from the pinion. Tighten, turning the drive shaft all the time to feel for resistance to the bearings getting tight.

When this happens tighten the adjuster ring firmly to make sure the outer races of the bearings are nicely seated against the adjusters. Slacken the adjuster ring again gradually to find where the preload comes off, and the gears are turned more easily. Bring in the adjuster again only this time to give the correct preload.

14 Fortunately the preload is such as to give a resistance to rotating the differential in its bearing of 1 lbf. ft. Make a little "balance" on the driving sleeve at the end of the shaft. Support the shaft so that it sticks out straight. Tie or bolt a stick or metal strip about 3 ft long across the drive sleeve. Half of it must stick out either side. Hang something weighing 1 lb on the cross bar a distance 1 ft from the centre. The bearings should be adjusted so that if they are one serration looser the weight is enough to rotate the differential.

15 Now adjust the backlash between crown wheel and pinion teeth. Slacken the right serrated bearing adjuster and tighten the left an equal amount to move the crown wheel closer to the pinion. The remaining backlash can be most easily felt at the pinion, by turning the end of the gearbox input shaft. It is easily moved but the crown wheel not, because of the bearing preload. Bring the crown wheel right in until all backlash dissappears.

16 Now back the crownwheel away again, continuing to move both bearing adjusters an equal amount. Try the size of the backlash after moving each serration.

17 To help judge the backlash, set one of the tappets on the engine to the backlash; 0.004 in compared with the tappet setting of 0.006 in. Compare the feel of these clearances. Try a tighter one.

18 Now get the backlash to this. It is better to be say 0.001 too tight rather than too loose, but no more.

19 All this movement of the bearings may have upset their preload. Check this again. This also should not be on the loose side or noise will develop when under load and the casing is hot. Check that the locking plates that fit under the oil retainers will fit into their slots.

20 All should now be set. So take off again from the gearbox casing the flywheel housing. Lift out the differential, and complete the reassembly of the gearbox.

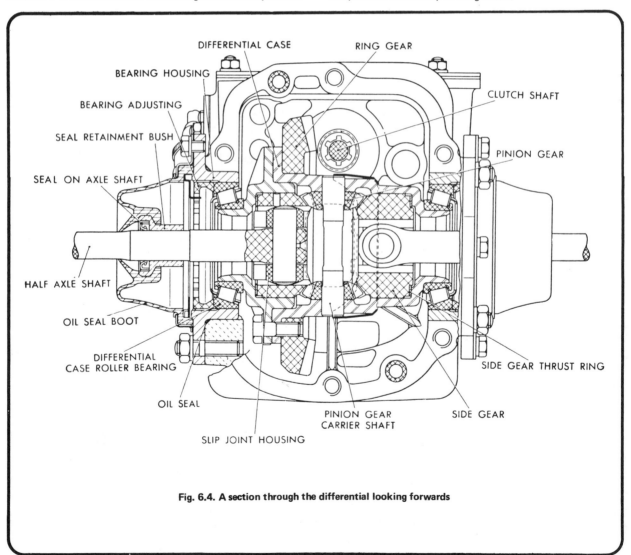

Fig. 6.4. A section through the differential looking forwards

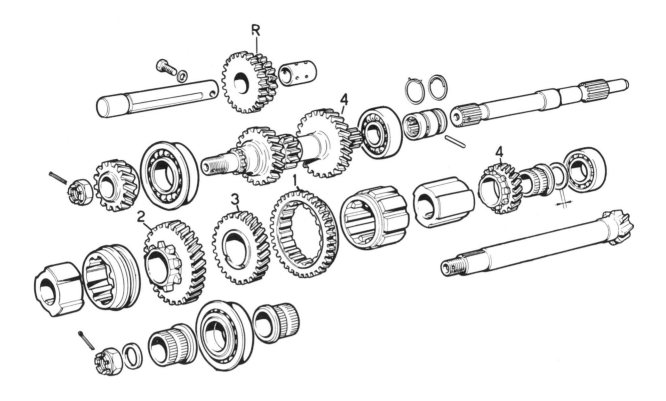

Fig. 6.5. Gear clusters and shafts with actual gear numbering

9 Reassembling the gearbox

1 Gather together all the parts. Make sure they and the casing are thoroughly clean. Have an oil can handy, filled with gearbox oil, 90EP. Assembly makes many oil drips, so put out some clean newspaper on the work surface.
2 All the parts want to be laid out in order. If they have been disturbed from the pattern in which they were laid out on stripping, have a trial assembly of everything onto the shafts, but not in the gearbox. Then you can compare everything with the pictures to check it is correct. Every part has some clue showing which way round it goes. At the worst you might have to take something out again to turn it round if it is wrong. Then lubricate all the parts.
3 Put the roller bearing on the output shaft, and slide it right the way along to the shoulder of the pinion. It may need driving along by a hammer. If so, support the inner race on something firm, such as the vice jaws opened enough not to touch the shaft, but to give good support, and cleaned, and hit the pinion.
4 Put the shims, if no parts changed the same as before, otherwise as calculated in sections 6/5 and 8/4, onto the shaft and close up to the bearing. Lubricate the shaft again.
5 Enter the threaded end of the shaft into its hole (the bottom one) in the casing from the differential end, so that it is just in far enough to start loading onto it all the bushes and gear wheels that must be fitted.
6 Fit the 4th gear bush, shouldered end first, and then the 4th gear wheel, its engaging-teeth-end last.

7 Put onto the output shaft the assembly of the first gear wheel, the 3rd/4th sliding dog-clutch sleeve, and the inner hub, engaging its keyway in the slot cut in the shaft. The groove for the 1st gear selector fork should be towards the differential end, but that on the sleeve for the 3rd/4th fork the other (selector end).
8 Last of this batch comes the 3rd gear bush, its shoulder last, with already on it the gear wheel, engaging dog-clutch teeth towards the differential.
9 All the parts that go on that output shaft within the main section of the casing should now be on the shaft.
10 Push the shaft on through the hole in the casing at the selector end, lining up the bearing by the pinion with its hole in the casing.
11 Fit the other bearing onto the shaft at the selector end, flange outermost, and slide it along the shaft to the casing. Now get both bearings entered into their holes in the casing. Once both are started, each work together to hold everything straight, so that the bearings can be pushed home.
12 Fit the ball bearing to the differential end of the input shaft, unless it stayed in the casing when stripped, and is still there. It seems easier to fit the bearing to the shaft outside, though it makes it a tight fit inserting the shaft with bearing into the casing. But if it came out on the shaft, then it will go back that way. Casings may differ.
13 Insert the input shaft into the casing, selector-end first, sliding it down so that the far end with the bearing on it comes down above the seat for it in the casing, where part of the casing has been machined away to allow just this. Manoeuvre it round

and down, and then forward again to enter the bearing in the differential-end of the casing.

14 Put the other bearing on the selector end of the input shaft, flange outermost, and the flat ground in that flange nearest the bearing on the output shaft, and slide it into its position in the casing.

15 Getting the shaft into its position lined up to fit these bearings will need some turning of the gear wheels down below on the output shaft to allow the gear teeth to engage.

16 With both shafts in position and their bearings fitted, put into place the housing for the output shaft bearing on the selector-end of the casing and fit the two countersunk screws, with conical lockwashers on cars after 1967, and tighten them.

17 Fit the bush for the 2nd gear to the output shaft, flange first, followed by its gear wheel, engaging dog-clutch teeth last.

18 Follow with the hub for the engaging sleeve, getting its key engaged in the groove on the shaft. (The sleeve will be fitted later).

19 Make sure all the parts on the end of the shaft are close up to the bearing housing, and give the shaft a push to make sure it is as far in as it will go. Then turn the shaft over by the pinion so that the seat, a little cup machined in the shaft, for the speedometer ball, is uppermost.

20 Fit the speedometer gear's driving ball in place in the cup, and then slide the speedometer gear wheel along the shaft, slot for the ball cut in the inside first, and get it into place.

21 Fit the washer and nut on the end of the shaft, tightening it finger tight for now.

22 To the input shaft fit the 2nd gear wheel, splined end first, followed by the nut (no washer). Again leave it finger tight for now. (Just in case something is wrong and it has to be taken apart again!)

23 Put the reverse gear shaft through its hole in the casing at the selector-end, its small end first, and then fit onto it inside the gearbox the reverse gear wheel, the smaller of its rings of teeth being towards the differential end.

24 Slide the reverse shaft on towards its location in the differential end of the casing, turning it so that the threaded hole in the end of the shaft is lined up for the bolt hole in the casing. Push it into place and insert the bolt.

25 Now check that nothing is wrong that will mean taking it apart again. Have you got the correct shims in for the pinion? The only gearbox parts left over should be the second gear sleeve, the selector forks and rods, and their detent and interlock balls. Engage each gear in turn, seeing that the hubs and wheels slide easily, and that the shafts turn easily.

26 Now tighten the retaining bolt for the reverse gear shaft.

27 Engage two gears to lock the gearbox, and then tighten the nuts on the ends of the two shafts, and fit their split pins. Unlock the gears, and check the shafts still turn easily.

28 Next the selector rods are fitted. Each can be recognised by the grooves cut at the "selector end" for the detents to hold the rods in the various positions, and the grooves, or hole in the middle (3/4th gears) one, for the interlock balls/plungers that prevent two gears being engaged simultaneously. See Fig. 6.5.

29 Enter the bottom, the 1st/2nd selector rod in the casing, this being the one with the second gear fork on its end combined with the actuating hook. As the rod is pushed into place, in the 2nd gear fork put the 2nd gear engaging sleeve and guide it onto the splined of its hub and within the fingers of the fork. Inside the gear box engage the first gear fork with its groove on the 1st gear wheel, and push the selector rod through it. The first gear fork is the stubby one. Line up the threaded hole in the rod with that in the fork, and insert its retaining bolt. (These are the special ones with long fat shanks). Tighten the bolt.

30 Insert a ball into the hole in the differential-end of the casing for the middle selector rod, and guide it down into the passage leading down to the lower rod. These "balls" are elongated, so must be fiddled around with a pair of screwdrivers till they stand on end to line up with the hole. Some grease makes them more amenable, as they will not fall over.

31 Get the middle rod ready. This is the one with a hole through its differential end. Into this hole put the thin plunger rod,

coating it with grease so that it is less likely to fall out.

32 Insert the middle rod in the casing, and slide it along till its selector fork, the 3rd/4th can be put on. The correct way round for the rod is to have the three detents at the selector end outwards, for the ball retainers to seat. Put the fork on the rod, slide the rod along into the differential end of the casing, and twist to line up the thread for the little retaining bolt for the fork. Fit this and tighten it. With the fingers on the forks different spans apart, the forks cannot be muddled, and they all go with their retaining bolts uppermost.

33 Insert the next "ball" into the hole in the casing at the differential end for the top rod, and again guide it into the downwards passage. Push and pull the middle and bottom rods so that they are in the neutral positions, and so that the balls and the plunger in the middle rod can sink to their bottom position, and not be in the way when the top rod is inserted.

34 Now fit the top, reverse, selector rod, with the long thin fork, guiding it in to line up the threads, fitting and tightening the bolt in the fork as for the others. Again the correct way round is shown by the detents at the selector end, and that at the other for the interlock, which must be downward.

35 Using the rods, select each gear in turn. Try to select two simultaneously to check the interlock.

36 Fit the selector-end cover. Grease its shaft, and put the gear selector lever a short way into its hole in the cover. Put the new gasket into place on the studs. Pick up the end cover and hook the end of the gear selector lever into position in the hooks on the end of the selector rods. Line up the holes in the cover with the studs on the end of the gear box casing. Push the cover onto the studs, maybe having to do so quite hard, to push the gear selector into the cover. Fit the spring washers and nuts, and tighten them evenly and diagonally.

37 Refit the speedometer drive, using a new paper gasket, and maybe giving it a twist to line up the gear teeth as it goes in. Fit the spring washer and nut.

38 Fit the top cover to the gearbox, using a new gasket, and tightening the screws evenly.

39 Fit the three balls and their springs into the side of the gearbox casing for the detents, and fit the cover, with new gasket.

10 Reassembly of the differential and drive shafts

1 In this section the reassembly is described assuming all parts are the old ones, or if new the meshing of crownwheel and pinion and the bearing preload has been reset.

2 Prepare the working area with layers of clean newspaper so that drops of oil will do no harm, and the parts stay clean. The oil can filled with 90 EP oil will be needed.

3 Oil all parts.

4 Fit the bronze thrust rings into each half of the casing ready for the side bevel gears.

5 Fit the cross heads to each drive shaft.

6 Put the right drive shaft into its bevel gear, getting the cross heads into the slot so that the bearing surface, the shiny one with the cuts for lubricant to move, taking the load.

7 Put that drive shaft and bevel gear into its half, the bigger one, of the differential case. Fit to the case the idler gears on their shaft, and lock them into place with the retaining cup.

8 Assemble the other drive shaft into its half of the casing with the other gear.

9 Fit the two halves of the casing together, lining up the bolt holes in the same position as before it was stripped.

10 Thread the crown wheel along the drive shaft and into place on the differential casing. Fit all the bolts with their lock washers, and then tighten them evenly and gradually.

11 Check that the gears inside the differential can be turned over by the drive shafts.

12 Fit the inner races of the roller bearings to both ends of the differential. Put the outer races into the bearing housings, and the new oil ring.

13 Fit the clutch shaft in its sleeve onto the gearbox input shaft.

Put in the pin, and lock that with its spring ring.

14 Fit the new oil seal for the clutch shaft to its hole in the flywheel housing. Tap it carefully in from the clutch side, with its lips towards the differential.

15 Lift the differential into place in the casing. If possible get an assistant to hold the gearbox upright so that the differential will sit in the bearing openings.

16 Fit the flywheel housing to the gearbox casing. Take care to slide it down the clutch shaft so that the oil seal is not strained. Make certain there is no dirt on the faces of the two halves of the casing, as there is no gasket, and they must seat in well.

17 Fit the star shaped bearing housings on either side, supporting the differential by the drive shafts to help the races get into place on the bearing rollers. Ensure that the housings are put back in their original places.

18 Fit the nuts to the studs holding the flywheel housing onto the gearbox casing, with their washers, and tighten them. Then fit and tighten the nuts for the bearing housings.

19 Check that everything can be turned over. First the clutch shaft, then one drive shaft alone: the other should turn backwards under differential action. Then turn both together. They will be stiff due to the bearing preload, but should turn smoothly.

20 Fit the locking rings to the serrated bearing adjusters, then the oil retaining boots, and their retaining rings, and bolt them into place.

21 Fit the dirt excluders to the outer ends of the drive shafts.

22 Smear their splines liberally with molybdenum-disulphide grease, and fit the driving sleeves.

23 Push on their circlips to the end of the drive shafts.

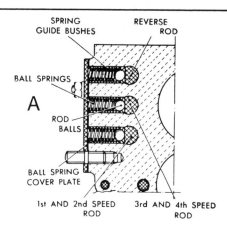

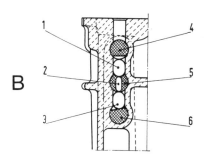

A Fig. 6.6. Spring loaded detent balls to hold gears in mesh or out

B Fig. 6.7. Inter-lock elongated balls (1 and 3) and plunger (2) to prevent simultaneous engagement of two gears. The top rod (4) engages reverse, the centre (5) third and fourth, and the bottom (6) first and second gears.

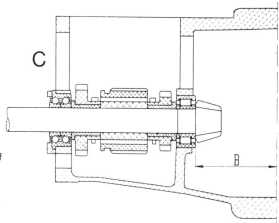

C Fig. 6.8. The dimension 'B' is the distance from the centre line of the differential to the rear face of the pinion. It is nominally 75 mm. The variation + or − of this that an individual pinion needs for correct meshing is stamped on the threaded end of the shaft

11 Refitting the transmission to the car

1 Normally the transmission would be refitted with the engine, and this is described in Chapter 1. This section deals with those few cases when it has been removed on its own, the vehicle being over a pit.

2 Fit the new mounting rubbers to the casing, and to them the crossmember.

3 It is assumed the engine is still sitting on a jack in the car, and that the clutch has not been disturbed, so its driven plate is still centralised.

4 Wipe any oil or grease off the splines on the clutch shaft. Smear a little grease on the very end of the shaft, where it goes in the spigot bearing in the crankshaft.

5 Get a jack or an assistant ready to hold the transmission in place whilst it is bolted up.

6 Lift the transmission into line with the engine. Make sure it is straight so that the clutch shaft can slide into the driven clutch plate without straining the clutch, or fouling the release mechanism.

7 Push it onto the engine. It may need a twist to get the splines on the shaft into the clutch plate.

8 Fit all the nuts holding the engine and transmission together, with their washers. Tighten them all evenly, supporting the transmission all the while.

9 Still supporting the transmission, fit the bolts holding the crossmember to the floor.

10 Remove the support under the transmission and the engine.

11 Refit the starter motor.

12 Reconnect all the controls, the starter motor, and the speedometer cable.

13 Connect up the drive shafts to the hubs, fitting in between the little springs.

14 Fill the transmission with oil.

15 Recheck all is refitted correctly.

16 Road test the car, and then check all bolts for tightness, and the transmission for leaks.

17 After 500 miles change the transmission oil to wash out any dirt that might inadvertently have got in during the work, and the metal rubbed off new parts.

12 Gear lever adjustment

1 If difficulty is found in engaging odd pairs or even pairs of gears the lever may need adjustment.
2 The gear lever mounting is fixed to the right side of the spine (or tunnel) down the centre of the floor by two bolts; with elongated holes to allow adjustment.
3 If 1st and 3rd gears are difficult to engage push the lever assembly forwards. Move it back if 2nd and 4th will not engage.

13 Smooth silent gear changing

1 If the gears are not engaged smoothly and correctly synchronised their engagement dog clutches will wear. Chips broken off them will grind away other components. After long abuse a gear may not work at all. (It usually happens to third gear first).
2 The double declutch method of gear change should be used. 500 gear box is very forgiving, and the gears will engage quietly with the speed mismatched to quite a wide extent.
3 If the mismatch of gear speeds is too wide, yet enough to caust actual geat grating, the transmission may produce a usual, and loud, clonk.
4 If there is difficulty engaging a gear silently from neutral, when moving off, even after waiting for a few moments after disengaging the clutch for the gear wheels to come to rest, then refer to clutch defects (drag) in Chapter 5.

14 Fault finding

1 Faults can be sharply divided into two main groups: Some definite failure with the transmission not working: Noises implying some component worn, damaged, or out of place.
2 The failures can usually be tracked down by commonsense and remembering the circumstances in which they appeared. Thus if the car will not go at all a mechanical failure will occur in different circumstances to a broken linkage from the gear lever!
3 If there is a definite fault within the transmission then it has got to be removed and dismantled to repair it, so further diagnosis can wait till the parts can be examined.
4 But if the problem is a strange noise the decision must be taken whether in the first place it is abnormal, and if so whether it warrants action.
5 Noises can be traced to a certain extent by doing the test sequence as follows:
6 Find the speed and type of driving that makes the noise. If the noise occurs with engine running, car stationary, clutch disengaged, gear engaged: The noise is not in the transmission. If it goes after the clutch is engaged in neutral, halted, it is the clutch.
7 If the noise can be heard faintly in neutral, clutch engaged, it is in the gearbox. It will presumably get worse on the move, especially in some particular gear.
8 Final drive noises are only heard on the move. They will only vary with speed and load, whatever gear is engaged.
9 Noise when pulling is likely to be either the adjustment of preload of the differential bearings, or the crown wheel and pinion backlash.
10 Gear noise when free-wheeling is likely to be the relative positions of crownwheel and pinion.
11 Noise on corners implies excessive tightness or excessive play of the bevel side gears or idler pinions in the differential.
12 In general, whining is gear teeth at the incorrect distance apart. Roaring or rushing or moaning is bearings. Thumping or grating noises suggest a link out of a gear tooth.
13 If subdued whining comes on gradually, there is a good chance the transmission will last a long time to come.
14 Whining or moaning appearing suddenly, or becoming loud, should be examined quickly.
15 If thumping, or grating noises appear stop at once. If bits of metal are loose inside, the whole transmission, including the casing, could quickly be wrecked.

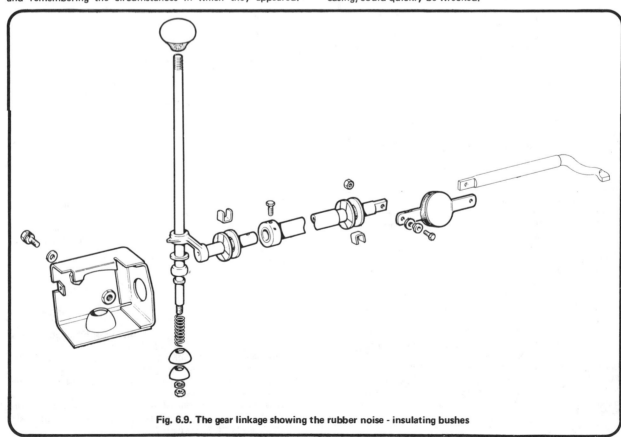

Fig. 6.9. The gear linkage showing the rubber noise - insulating bushes

Chapter 7 Braking system

Contents

General description ... 1	Wheel operating cylinder - replacement ... 11
Safety maintenance ... 2	Hydraulic master cylinder - removal and overhaul ... 12
Routine maintenance ... 3	Handbrake - adjustment... 13
Bleeding the brakes ... 4	Handbrake cable - replacement... 14
Flexible hose inspection - removal and replacement... 5	Brake backplate - removal ... 15
Brake adjustment - (station wagon) ... 6	Self adjuster problems ... 16
Removing front brake drums ... 7	Oval or scored drums ... 17
Removing rear brake drums ... 8	Brake fade.. ... 18
Replacement of brake shoes ... 9	Fault diagnosis and remedies ... 19
Wheel operating cylinder - overhaul in place ... 10	

Specifications

Sedan brake drum: Internal diameter ... 6.702 - 6.712 inch (170.23 - 170.48 mm)
Station wagon: Drum diameter ... 7.293 - 7.304 inch (185.24 - 185.53 mm)
    Maximum permissible machining ... 0 - 039 inch (1.0 mm)

Brake lining
    Minimum lining thickness ... .060 inch (1.5 mm)
    Lining bonding to shoe ... "Permafuse"
    Lining width ... 30 mm (1.181 inch)
    Lining length ... 180 mm (7.09 inch)
    Lining new thickness ... 4.2 - 4.5 mm (.165 - .177 inch)

Brake fluid
    Capacity ... Approx 0.2 qrt or litre
    FIAT "Azzura" (light blue) or to specification SAE J 1703b
    Castrol Girling Brake Fluid:
    Rod to piston clearance ... 0.5 mm (.020 inch)
    Pedal free travel ... 2.5 mm (.100 inch)
    Master cylinder diameter ... ¾ inch
    Wheel cylinder diameter ... ¾ inch

Tightening torques
    Nuts on bolts securing brake backplates to hubs:
    Front brakes ... 14½ lbf.ft (2.0 kg.m)
    Rear brakes ... 43 lbf.ft (6.0 kg.m)

1 General description

The FIAT 500 is fitted with drum brakes to the front and rear wheels. The station wagon has larger brakes than the sedan.

The brakes are of the internal expanding drum type and are operated by the master cylinder mounted on the front bulkhead. Front and rear are similar.

The brakes are of the single leading shoe type with one double ended wheel cylinder operating the two shoes. Attached to the rear brake assemblies is a mechanical expander operated by the handbrake lever through a cable which runs from the brake lever to the brake backplate levers. This provides an independent means of rear brake application.

On station wagons the brakes have to be adjusted periodically to compensate for wear in the linings. Self adjusting brakes are fitted as standard to sedans.

The need to adjust the handbrake is infrequent as its efficiency is largely dependent on the condition of the brake linings and the adjustment of the brake shoes. The handbrake can, however, be adjusted separately to the footbrake.

2 Safety maintenance

1 It is important that the complete braking system is at the peak of condition at all times. To safeguard against any weak or deterioration the system should be checked as follows, in addition to routine maintenance.
2 The brake hydraulic fluid should be completely changed every 36,000 miles (60,000 km). It attracts damp; so deteriorates.
3 All seals in the brake master cylinder and wheel cylinders and the flexible hoses should be examined and preferably renewed every 36,000 miles (60,000 km). The working surfaces of the master cylinder and wheel cylinders should be inspected for signs of wear, pitting or scoring and new parts fitted as considered necessary.
4 Only the recommended brake fluid should be used in the hydraulic system. Never leave brake hydraulic fluid in open unsealed containers as it absorbs moisture from the atmosphere which lowers the safe operating temperature of the fluid. Also any fluid drained or used for bleeding the system should be discarded.
5 Any work performed on the hydraulic system must be done under conditions of extreme care and cleanliness, and kept clear of ordinary oil, grease, petrol or paraffin.

3 Routine maintenance

1 The routine maintenance wear was all detailed at the front of the book. It must be done thoroughly.
2 In particular the removal of brake drums called for in the 6,000 mile task is an important task, and, is not over maintenance. The visual check is necessary for safety. Also all dust must be removed to ensure smooth braking.

4 Bleeding the brakes

1 Unless all air is removed from the hydraulic brakes their operation will be spongy, and pedal travel too long. Air will get into the system when it is dismantled or when there is a leak. Sometimes it can leak in when the fault is so slight no fluid will escape to make it apparent. Air could get into the system when the hydraulic fluid level became to low in the reservoir and air was pumped in by using the pedal.
2 To bleed the brakes you will need:
a) An assistant to pump the pedal.
b) A supply of new hydraulic fluid.
c) An empty glass jar.
d) A plastic, or rubber pipe, about 12 in (30 cm) long which will fit tightly over the bleed nipple on the operating cylinder.
e) A spanner to fit the nipple.
3 Fill up the master cylinder full.
4 Put fluid in the bleed jar to a depth of about ½ inch.
5 Each brake must now be bled in turn, starting with the wheel cylinder that is furthest away from the master cylinder.
6 Clean the bleed nipple. Put the pipe on the nipple. Tell your assistant to give a few quick strokes to pump up pressure on the pedal and then hold it on.
7 Slacken the nipple about ½ or 1 turn, until the fluid or air begins to come out. This is usually quite apparent either as bubbles or dirt in the clean fluid in the jar.
8 As soon as the flow starts tell the assistant to keep pumping the pedal every time it gets to the end of its travel, and to tell you all the time what he is doing: For example "down, up" etc.
9 As soon as air stops coming out of the bleed pipe in the jar shut the bleed nipple: Do so as the assistant is pushing the pedal downwards. Do not go on too long lest the reservoir be emptied, and more air pumped in. About 15 pedal pumps is safe.
10 Repeat at the other wheels. By this time all the fluid coming out of the nipples should be clean: If it is still dirty do some extra pumping to ensure that the opportunity has been taken to change the fluid.

11 The bleeding operation is made much simpler, and is possible alone, if to the bleed nipple holes in the operating cylinders are fitted automatic brake spring loaded bleed valves. These are available from accessory shops.
12 Keep the hydraulic fluid clear of the paint: It ruins it. Throw away the old fluid immediately after bleeding. It is unuseable anyway as it will be aerated. Also fluid absorbs damp, and so deteriorates in use.
13 Use this bleeding sequence when changing the brake fluid at 36,000 miles.
14 It may be found that the bleed operation for one or more cylinders is taking a considerable time and the cause is probably due to air being drawn past the bleed screw threads when the screw is loose. To counteract this condition, at the end of each downward stroke tighten the bleed screw to stop air being drawn past the threads; then open as the pedal is pressed again.
15 If after the bleed operation has been completed the brake pedal operation still feels spongy, this is an indication that there is still air in the system, or that the master cylinder is faulty, and sucking air in.

5 Flexible hose inspection, removal and replacement

1 Inspect the condition of the flexible hoses in the brake hydraulic system for signs of swelling, cracking, cuts, or chafing, and if any are evident the hose must be replaced.
2 Wipe the top of the fluid reservoir and unscrew the cap. The fluid must be plugged into the reservoir to prevent it flooding out when the pipe is disconnected. One way is to put a piece of plastic sheet over the top, and replace the cap to hold it. This allows a little initial loss. A surer way is to put a wooden plug such as a pencil, though sharpened to a narrower taper than normal, into the outlet at the bottom of the reservoir. It must be short so that the cap can be put on again to keep out the dirt.
3 Clean the area around the ends of the pipe.
4 Unscrew the metal pipe union from its connection (at one end of the pipe for the front, and both ends of that at the rear brakes). Then holding the hexagon on the hose with one spanner to prevent it turning, undo the nut holding the flexible pipe to the bracket.
5 On the front brakes undo the union nut on the back of the wheel cylinder. Note that there are washers either side of the union.
6 When reassembling make sure everything is clean, especially the washers for the union at the front.
7 Unplug the reservoir. Top it up, and then bleed the brakes as described in section 4.

6 Brake adjustment - (station wagon)

1 The brake shoes each have a snail cam adjuster.
2 Before adjusting the drum to shoe lining clearance, the brakes should be firmly applied several times so as to centre the shoes.
3 Jack up one wheel.
4 An assistant should now depress the brake pedal so that the brake shoe linings are in contact with the brake drum. Keep the brake pedal firmly depressed and using a spanner turn the adjuster cam outwards as far as it will go, but without exerting excessive force.
5 Now turn back the adjuster by about 25° for new linings or 20° for used linings. This will give approximately 0.01 inch (0.25 mm) clearance at the shoe. Repeat this procedure for the second cam adjuster.
6 Release the brake pedal and make sure that the wheel can be rotated freely. NOTE: a rubbing noise when the wheel is rotated is usually due to dust on the brake drum and shoe lining. If there is no obvious slowing of the wheel due to brake binding, there is no need to back off the cam further until the noise disappears. It is better to remove the drum and re-dust, taking care not to inhale any dust as it is of an asbestos nature.
7 Repeat this sequence for the remaining wheels.

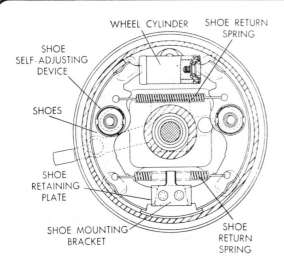

Fig. 7.1. Front brake parts (left side)

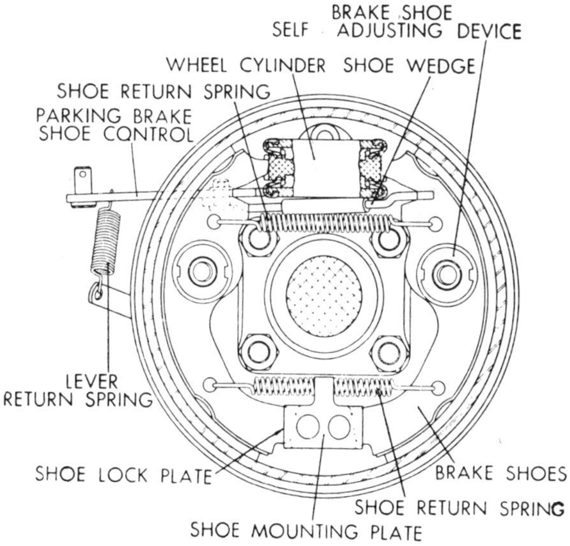

Fig. 7.2. Rear brake parts (right side)

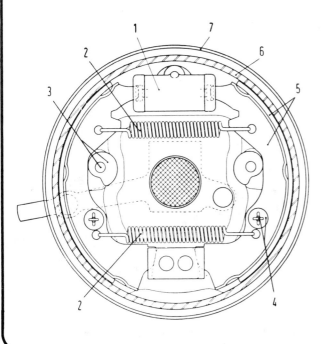

Fig. 7.3. Station wagon brakes (front left)

1 Wheel cylinder
2 Pull-off springs
3 Adjusting cams
4 Shoe steady pins
5 Shoe and lining
6 Drum
7 Back plate

Chapter 7/Braking system

7 Removing the front brake drums

1 The front brake drum and hub are all made in one assembly. To take off the drum the hub bearing must be dismantled.
2 Jack up the wheel of the car, and undo the bolts retaining the wheel and remove it.
3 Clean the outside of the brake drum and the boss and cover where the housing is for the wheel bearings.
4 Prise off the cover to the bearing.
5 Undo the wheel bearing nut. On early cars this had a split pin and castellated nut. All later cars have self locking nuts, which are held in place by being peened into a slot on the stub axle. Hammer out this peening with a punch. Undo the nut: Note that on the right hand side it is a left hand thread. Now pull off the brake drum and hub complete. If the bearing will not slide off the stub axle by hand, use a puller. A conventional puller will have to be rather large to reach out to the lugs for the wheel nuts. It is possible to make one using the wheel. Replace the wheel on the drum, taking care not to scatter any dirt inside the bearing. Fit a long bolt through the threaded hole in the strap over the wheel that holds the hub cap on. Screw in this bolt to withdraw the hub.
6 Before reassembling wipe away the old grease at both ends of the bearing, taking care not to damage the seal. Repack with clean grease. Do not fill the hub tightly with it, but merely smear the bearings liberally with it. Clean the stub axle and lightly smear it with grease. Put the hub carefully back on the stub axle, taking care not to damage the grease seal as its threaded over. Replace the washer, and then put on the nut.
7 Tighten up the bearing nut. If these nuts are the early type, castellated and using a split pin, they should be tightened to a torque of 22 lb ft (3 kg m), then undo 60° (1 flat of the nut) and the split pin inserted. The more recent self locking nuts should be tightened to a torque of only 5.1 lb ft (0.7 kg m). They should then be backed off 30° (½ a flat of the nut) and then locked by peening in a soft dome of the nut into the groove on the stub axle. It is likely that new nuts will be necessary due to the old mark on an old nut. It is recommended that new nuts are used each time the front hub is dismantled.

8 Removing the rear brake drums

1 Chock the wheels of the car on the other side, and release the handbrake.
2 Remove a rear wheel.
3 Undo the four bolts holding the drum to the hub.
4 Pull off the brake drum. It may be rusted in place. If so squirt with a rust remover, such as an aerosol WD 40. By hammering the drum where it sits on the flange of the hub, the rust can be jogged out of place, and the drum bounced slightly in place to free it.
5 Before reassembling wipe a slight trace of grease on the hub flange to make it easier to remove next time. (Too much grease will get flung onto the brakes).

9 Replacement of brake shoes

1 If the brake shoes are worn down to the condemnation limit of 1.5 mm (.060 in), or have been badly contaminated with grease or brake fluid, they must be replaced. The linings are bonded to the shoe, so it is not practical for an amateur to reline the old shoes. Buy the shoes from the official FIAT source, as wrong linings can be lethal. It is important that whatever is done to one side is repeated on the other so that uneven braking is avoided. So if one side is replaced the other must be too, even if otherwise not necessary. It will be found that the leading shoes wear more rapidly than the trailing ones. It is satisfactory to fit new linings just to the leading shoes.
2 The front and rear brakes are the same except for the hand brake linkage. The station wagon does not have the self-adjusters. Note the position of the pull-off springs before you

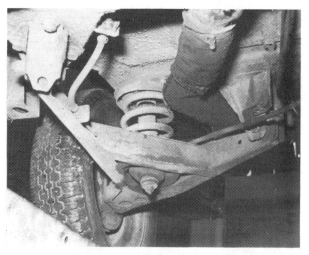

Fig. 7.4. The brake flexible pipes at the rear are inboard of the suspension pivot

Fig. 7.5. The shoes must be disengaged from the piston at the top, the pivot at the bottom, and the self adjuster stud

Fig. 7.6. The self adjusters are difficult to undo without special spanners. You can make them out of tubing

start. Both should be hooked into the shoes from in front. Remove the drums as described in the previous two sections.

3 Removal of the shoes is somewhat impeded by the self adjusters. To take the shoe off, pull the top end outwards, so that it is disengaged from the groove on the piston in the hydraulic cylinder. Then lever it towards you slightly, and let it go back again relieving the spring tension, but not clear of the piston in the operating cylinder. Then do the same at the fixed pivot at the bottom. Now wriggle the shoe off the pivot stud for the self adjuster. It is assumed that if you are right handed you are doing the right hand shoe as you look at the assembly. The second shoe comes off easily afterwards.

4 If great difficulty is experienced doing it as described, then unhook the brake, pull off springs on the shoes. To do this push them with a screwdriver. This requires less knack, but is more likely to result in bruised or cut knuckles.

5 On the rear brakes as the shoes come off, note how they are fitted into the linkage for the handbrake.

6 With the shoes removed clean the brake backplate assembly. Check that the securing bolts are tight.

7 If new shoes are to be fitted the self adjusters must now be taken off the old ones and transferred. The two halves of the self adjuster are threaded together, clamping the large coil spring between them. Special spanners are normally used to undo these. As the flanges are rather narrow it is difficult to do so without them. Also the self adjusters are made of sintered metal, and brittle. It is suggested that as you will be getting new shoes from your FIAT agent the opportunity is taken to get them to transfer the self adjusters over. Otherwise special spanners will need to be made, and taken to a blacksmith for hardening.

8 If you succeed in undoing the self adjusters yourself, note that there are friction washers next to the shoe on either side.

9 Before reassembling the shoes very lightly smear with grease the area of the shoes where the self adjusters go, and their pivots for the fixed mounting at the bottom. But do not put any grease on the end of the shoe going on the hydraulic cylinder.

10 Take the opportunity to check that hydraulic fluid is not seeping from the dirt excluders on the operating cylinder.

11 To reassemble the shoes to the brake, first put in place the left hand shoe (assuming you are right handed) with both pull off springs hooked in position on that shoe. Now bring the other shoe close in until you can hook on the pull off springs into the holes. To do this both ends of the shoe must be outside the places where they will normally rest on the operating cylinder at the top and the pivot at the bottom.

12 Having hooked on the springs pull the shoe outwards to hold them tight, and to line up the self adjuster with its stud on the brake backplate. Now wriggle the shoe down the adjuster, and get the bottom pivot onto its fixed abutment on the backplate, and then the other end onto its slot on the operating cylinder piston.

13 Having got the shoes into place push them back, compressing the operating cylinder pistons, so that the brakes will be well away from the drums to make these easy to fit.

14 Having reassembled all the brakes apply the brake pedal a number of times to adjust them on the self adjuster.

15 On the station wagon small steady springs hold the shoes to the backplates. Put a finger behind the brake backplate on the head of the pin. With pliers, push the washer on the outside of the steady spring towards the brake shoe, and turn it through 90° to release the pin head.

16 Adjust the brakes on the station wagon when new shoes have been fitted, and readjust them after 500 miles, when the linings will have bedded in.

10 Wheel operating cylinder - overhaul in place

1 The operating cylinders at the wheels must be overhauled if there are hydraulic leaks, or if air keeps getting into the system. These are more likely to be at fault than the master cylinder, because they operate in a harsher environment.

2 Jack up the wheel, and remove the brake drum as previously described.

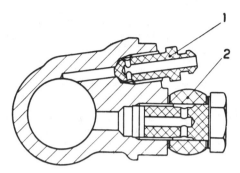

Fig. 7.7. Section of front wheel operating cylinder

1 Bleeder screw
2 Fluid pipe union with washers either side

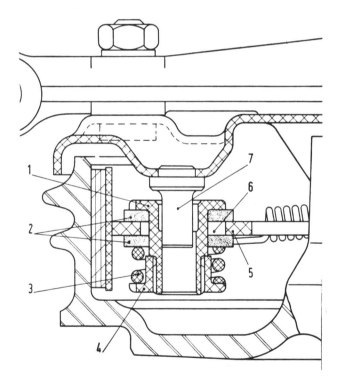

Fig. 7.8. Section view of the self adjuster

1 Adjuster body
2 Friction washers
3 Spring
4 Bush top
5 Brake shoe
6 Slot in shoe
7 Stud on back plate

3 Plug the outlet pipe from the hydraulic supply tank with a thin wooden taper such as a pencil as described earlier.
4 But if the brake fluid is old and dirty, take this opportunity to drain it all away.
5 Remove the brake shoes as described in section 9.
6 Wash the wheel cylinder outside with brake fluid.
7 Take off the rubber boots from either end of the cylinder, and push out the pistons and all rubbers. Note which way round they were for reassembly. See also the attached diagram. Clean out the cylinder. If the cylinder is pitted or scored leaks will recurr, and a replacement unit should be fitted.
8 In reassembling the brakes wash all parts and lubricate them with brake fluid. Use new rubber cups and rubber dirt excluding boots. These come as an overhaul kit. Treat the cups very carefully, handling them only with the fingers. They go in with their sharper face innermost. Take care the lip does not get damaged as it enters the cylinder.
9 Unplug the hydraulic supply tank. Pump up pressure in the brakes with the pedal, and refill the hydraulic supply.
10 Bleed the brakes as described in section 4.

11 Wheel operating cylinder - replacement

1 If the walls of the wheel operating cylinder have been pitted by corrosion, or scored by dirt, then a replacement unit must be fitted. If this is not done replacement cups will only last a short time, and recurring fluid leaks will contaminate the linings.
2 Jack up the wheel, and remove the brake drums as previously described.
3 Seal the hydraulic supply tank either by putting plastic under the lid, or by inserting a thin wooden taper such as a pencil into the outlet union at the bottom. But if the brake fluid is old, take this opportunity to drain it all out.
4 Clean all around the operating cylinder both in front and behind the backplate, and the hydraulic pipe union into the cylinder.
5 Disengage the brake shoes from the operating cylinder pistons. It may be possible to lever them back, and the self adjusters hold them clear, or it may be possible to wedge them back with chocks. Push the pistons into the cylinders to clear the ends of the shoes. If this fails, then the shoes must be removed as described in section 9.

6 Disconnect the hydraulic pipe from the back of the cylinder. On the front brakes there is a large union nut, with a washer either side. At the rear it is a simple union.
7 Now undo the two screws on the brake backplate holding the operating cylinder. Then lift the cylinder clear.
8 Clean the back plate where the cylinder sits, then fit the new one. Ensure cleanliness in the unions. Do any washing with brake fluid.
9 Connect up the pipe.
10 Release or refit the brake shoes.
11 Unplug or refill the fluid reservoir.
12 Pump the brakes a few times to work up pressure. Check for leaks. Recheck the fluid level.
13 Bleed the brakes as described in section 4.

12 Hydraulic master cylinder - removal and overhaul

1 Provided correct brake fluid is always used, and provided the brake fluid is changed as described at intervals of approximately 36,000 miles, no trouble should be experienced with the master cylinder. However a previous owner might have involved you in trouble. The faulty operation of the master cylinder is likely to

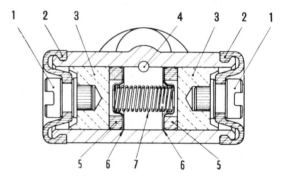

Fig. 7.9. Wheel cylinder parts

1 Shoe seats
2 Rubber boots
3 Pistons
4 Fluid inlet
5 Cups
6 Spring seats
7 Spring

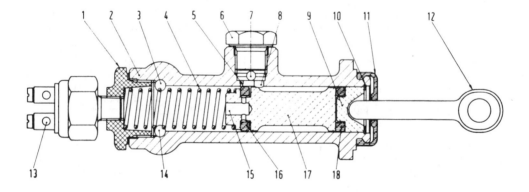

Fig.7.10 Master cylinder parts

1 End plug
2 Body
3 Outlet to front brakes
4 Piston return spring
5 Compensating hole
6 Plug
7 Inlet passages
8 Inlet passages
9 Plunger
10 Circlip
11 Boot
12 Push rod
13 Brake light switch
14 Outlet to rear brakes
15 Return hole in piston
16 Main cup
17 Piston
18 Secondary cup

Chapter 7/Braking system

show itself either in leaks inside the car from the end of the cylinder down by the pedal, or outside between the mounting flange for the master cylinder and the car body. Also air might leak into the system, without actually showing a leak by the loss of fluid: It is possible for a minor defect to allow air in but not fluid out. This would mean that the brakes would frequently need bleeding, due to them becoming spongy, despite the wheel operating cylinders being in good condition. If the wrong fluid is used the rubbers may swell and prevent proper piston movement.

2 The master cylinder is removed as follows:

3 Blank off the brake fluid reservoir, to prevent loss of fluid whilst the piping is disconnected. Do this by making a small, gently tapering plug, to put in the outlet at the bottom of the reservoir. A pencil will make an ideal plug, though it needs to be sharpened to a more gradual taper than is normal. Or put a sheet of plastic over the top, and replace the lid.

4 Disconnect the two pipes to the two front brakes, and the pipe to the rear brakes from the master cylinder. Disconnect the pipe from the fluid reservoir into the master cylinder. Unclip the wires for the brake light switch on the front.

5 Undo the two nuts holding the master cylinder to the car body and lift it clear.

6 Clean the master cylinder carefully on the outside, using either dry rag or brake fluid.

7 Take off the rubber boot from the end of the cylinder, the two plugs in the cylinder, and the brake light switch.

8 Take out the circlip inside the cylinder.

9 Take out the plunger, and the piston with all rubber rings, washers, and springs. See Fig. 7.8. Note the order and way round in which they were in the cylinder. To get them out it may be necessary to push with a long thin and clean screwdriver through the hole where the brake light switch was.

11 Clean all parts in brake fluid.

12 Reassemble using all new rubbers. These can be supplied in a complete kit. Take great care fitting them. The sharper lips of the cups go inwards. Only use fingers to get them on the piston.

13 Clean the car body where the master cylinder is mounted.

14 Reassemble the master cylinder to the car, and reconnect all pipes and the brake light wiring.

15 Remove the plug in the reservoir. Pump up pressure in the system on the pedal, and then refill the hydraulic supply.

16 Bleed the brakes as described in section 4.

17 Check there is 0.5 mm (.020 in) clearance between the rod worked by the pedal and the plunger. This at the pedal giving 2.5 mm (.100 in) free play. The free play is not normally adjusted. It will become wrong if the pedal bracket is misplaced.

Fig. 7.11. Master cylinder from below, with steering box removed just for the photo. The inlet pipe is nearest the bulkhead. The other pipes go to front and rear brakes, and the brake light switch is on the end

13 Handbrake - adjustment

1 The handbrake is cable operated. One cable goes from the lever to both sides, it being connected to the lever by passing round a wheel which equalizes the effort to left and right.

2 The handbrake cable has adjusters at either rear wheel fitting. Adjustment for wear of the brake shoes will be taken care of by the self adjusters within the drum. Adjustment of the cable is only needed to take care of the stretch and wear in its links.

3 If handbrake movement becomes excessive, that is moves more than four clicks of the ratchet, proceed as follows

4 Drive the car on the road both forwards and reverse, applying the brakes by the brake pedal quite sharply, to make sure that the self adjustment has been fully taken up within the drum.

5 Jack up the rear of the car. Check that the wheels turn quite freely, without any brake binding.

6 Apply the handbrake two clicks up the ratchet. Now tighten the cable adjusters on both sides of the car by the same amount, to keep the bend in the cable in the same position on the wheel of the handbrake. Tighten them until a definite rubbing of the shoe on the brake drum can be felt, and the wheels can only just be turned.

7 Now release the handbrake and check the wheels are quite free.

8 Road test. After stopping uphill by the gears, check the brake drums are quite cold.

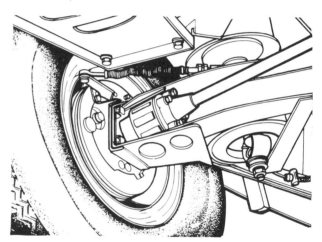

Fig. 7.12. The station wagon rear brake showing the shoe adjusters, and the hand brake, the latter being the same as the sedan.

14 Handbrake cable - replacement

1 If the handbrake cable should break, or even show signs of age with fraying, it will need changing.

2 Remove the split pin from the pin holding each cable end fork to the lever on the handbrake mechanism on the rear of the brake drums.

3 Having pulled out the pins from the forks undo the adjuster nuts and lock nut from the bracket on the rear suspension swinging arm. This is a 'U' shaped bracket. The cable can be pulled along until the inner cable can pass through the slot in the bracket.

4 Inside the car remove the two small bolts holding the tin cover to the panel just in front of the rear seat.

5 Undo the bolts holding the handbrake bracket to the top of the tunnel. Now pull up the handbrake assembly until the pin through the wheel working the cable can be reached. Take out the split pin, remove the wheel's pin, and disengage the cable from the lever.

6 Disengage the ends of the outer cables from their seats in the holes in the rear bulkhead. Slide them along to the larger holes nearby. Then the cable can be threaded out of the car.
7 When fitting the new cable grease all the pivot pins.
8 Adjust the cable as described above.

15 Brake back plate - removal

1 The brake back plate is likely to need removal for work on the suspension; particularly is this so at the front if the king pins need attention.
2 Jack up the car and remove the appropriate wheel.
3 Take off the brake drum as described in sections 7 for the front and 8 for the rear.
4 Plug the outlet in the fluid reservoir with a tapering wooden plug, and refit its lid to keep out dirt.
5 Clean the area around the hydraulic pipe union to the wheel operating cylinder on the back of the back plate. Undo the pipe union from the cylinder. Take care of the two large washers on the front brakes.
6 For the rear brakes the hub flange will now need to be taken out as described in Chapter 9.2.
7 Undo the four bolts holding the rear brake's back plate to its mounting on the suspension arm.
8 On the front brakes, undo the nut on the bolt tying the back plate to the steering arm, and the other just behind the king pin.
9 When reassembling make sure the seating for the back plate is thoroughly clean.
10 On many occasions the front brake assembly on the back plate can be removed but the flexible pipe need not be disconnected. So the labour of bleeding is avoided. In this case merely unbolt the back plate, then taking care not to strain the flexible pipe, hang the back plate up with a bit of string, out of the way.

16 Self adjuster - problems

1 In normal running the self adjusters should keep the pedal travel constant.
2 If the travel becomes long the immediate reaction is that the linings are worn out, and the adjusters have reached the end of their travel. However this condition should not arise if the proper examination of the brakes has been made, as the thin linings would be seen and changed earlier during routine maintenance.
3 If the linings are in order but the travel is large it can sometimes be due to the adjuster for the trailing shoes not being able to get that one close enough, as the trailing shoe is not worked hard. This can be helped by driving in reverse, and applying the brakes several times fairly hard.
4 After extended mileages the self adjusters themselves get worn. The clearance between the stud on the back plate and the hole in the self adjuster is designed to allow the brakes to come off the required amount when new. It therefore pays when fitting new brake shoes on older cars to fit new self adjusters as well on the assumption that the old ones are the originals. This should not be done at every renewal of brake linings.

17 Oval or scored drums

1 The drums wear with use, and this wear is often uneven.
2 Ovality of the drums makes itself apparent by juddering at speed. Other things also cause juddering, but ovality can also be felt at slow speed as a rythmic reaction at the pedal. The ovality can actually be felt.
3 Scoring gives rapid lining wear.
4 Some scoring very soon comes to the drums working surface. When excessive the drums can be machined a little, which should be enough to restore a smooth surface. As a professional will be needed to do the turning, he can advise as to whether the scoring is serious enough to warrant it.

18 Brake fade

1 If brakes get very hot the coefficient of friction of the brake linings, that is their grip of the drum, is lowered. Thus the braking power is reduced. This is aggravated as the wrapping effect of the leading shoe is reduced too.
2 It would need very hard use of the brakes to give fade on such a light car. But it could happen when well laden and descending a long hill.
3 If brake fade is experienced it will be found that the stopping power is restored as soon as the brakes have had but quite a short opportunity to cool a little from the critical temperature.
4 The rule for avoiding brake fade is to avoid working them hard on long descents, by going down a hill in the same gear as would be used to climb it.

Fig. 7.13. The routing for the handbrake cable in the spine/tunnel at the back

Fig. 7.14. As well as the one behind the king pin, there is a nut on the steering arm holding the back plate of the front brakes

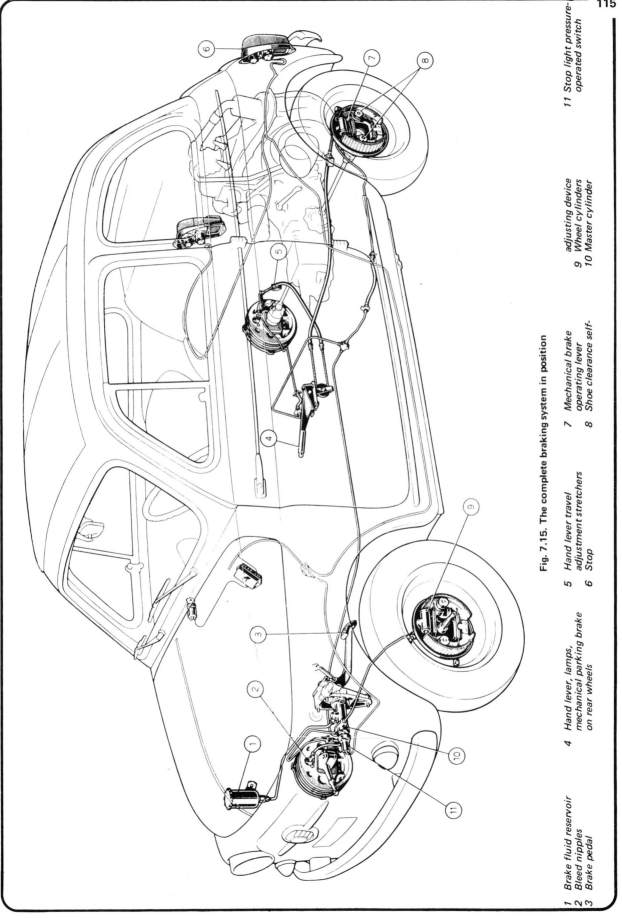

Fig. 7.15. The complete braking system in position

1 Brake fluid reservoir
2 Bleed nipples
3 Brake pedal
4 Hand lever, lamps, mechanical parking brake on rear wheels
5 Hand lever travel adjustment stretchers
6 Stop
7 Mechanical brake operating lever
8 Shoe clearance self-adjusting device
9 Wheel cylinders
10 Master cylinder
11 Stop light pressure-operated switch

19 Fault diagnosis and remedies

Before diagnosing faults in the brake system check that any irregularities are not caused by:

1. Uneven and incorrect tyre pressures
2. Incorrect 'mix' of radial and crossply tyres
3. Wear in the steering mechanism
4. Defects in the suspension and dampers
5. Misalignment of the bodyframe

Symptom	Reason/s	Remedy
Pedal travels a long way before the brakes operate	Brake shoes set too far from the drums	Adjust the brake shoes to the drums. (This applies equally where disc brakes are fitted but only the rear drums need adjustment).
Stopping ability poor, even though pedal pressure is firm	Linings and/or drums badly worn or scored	Dismantle, inspect and renew as required.
	One or more wheel hydraulic cylinders seized, resulting in some brake shoes not pressing against the drums (or pads against discs)	Dismantle and inspect wheel cylinders. Renew as necessary.
	Brake linings contaminated with oil	Renew linings and repair source of oil contamination.
	Wrong type of linings fitted	Verify type of material which is correct for the car and fit it.
	Brake shoes wrongly assembled	Check for correct assembly.
Car veers to one side when the brakes are applied	Brake linings on one side are contaminated with oil	Renew linings and stop oil leak.
	Hydraulic wheel cylinder(s) on one side partially or fully seized	Inspect wheel cylinders for correct operation and renew as necessary.
	A mixture of lining materials fitted between sides	Standardise on types of linings fitted.
	Unequal wear between sides caused by partially seized wheel cylinders	Check wheel cylinders and renew linings and drums as required.
Pedal feels spongy when the brakes are applied	Air is present in the hydraulic system	Bleed the hydraulic system and check for any signs of leakage.
Pedal feels springy when the brakes are applied	Brake linings not bedded into the drums (after fitting new ones)	Allow time for new linings to bed in after which it will certainly be necessary to adjust the shoes to the drums as pedal travel will have increased.
	Master cylinder or brake backplate mounting bolts loose	Retighten mounting bolts.
	Severe wear in brake drums causing distortion when brakes are applied	Renew drums and linings.
Pedal travels right down with little or no resistance and brakes are virtually non-operative	Leak in hydraulic system resulting in lack of pressure for operating wheel cylinders	Examine the whole of the hydraulic system and locate and repair source of leaks. Test after repairing each and every leak source.
	If no signs of leakage are apparent the master cylinder internal seals are failing to sustain pressure	Overhaul master cylinder. If indications are that seals have failed for reasons other than wear all the wheel cylinder seals should be checked also and the system completely replenished with the correct fluid.
Binding, overheating	Master cylinder faulty cups	Overhaul master cylinder.
	Master cylinder no free play	Check pedal clearance.
	Handbrake too tight	Re-adjust.
	Station wagon: adjusters too tight	Re-adjust.
Vibration, pedal pushed up in place with slow vibration	Drums worn oval	Have drums skimmed.
Juddering	Loose brake backplate	Tighten.
	Dust in drums, or oily linings	Clean and/or reline.

Chapter 8 Electrical system

Contents

General description ... 1	Starter motor - description ... 16
Battery - inspection ... 2	Removing and refitting the starter motor ... 17
Battery electrolyte ... 3	Stripping the starter motor ... 18
Battery - charging ... 4	Checking the starter motor ... 19
Battery leaks or corrosion ... 5	Reassembling the starter ... 20
Dynamo - description ... 6	Fuses ... 21
Generator - fault finding.. ... 7	Windscreen wiper... ... 22
Dynamo - removal and replacement ... 8	Horn ... 23
Dynamo - stripping ... 9	Lighting ... 24
Dynamo - overhaul ... 10	Headlight - aiming ... 25
Dynamo - reassembly ... 11	Headlight - changing ... 26
Control box - function and faults ... 12	Fitting accessories ... 27
Cutout checks ... 13	Instruments ... 28
Voltage regulator checks ... 14	Fault finding principles ... 29
Current regulator checks ... 15	Fault finding table ... 30

Specifications

Battery

Voltage: nominal ...	12 volts
Capacity (at 20 hour rate) ...	32 amp. hrs
Length...	235 mm
Width ...	133 mm
Height...	198 mm
Electrolyte level ...	3 mm (1/8 inch) above separators
	Or to filler bottom on special batteries
Specific gravity: Fully charged ...	1.272 at 32°C
	1.276 at 27°C
	1.280 at 21°C
	1.284 at 16°C
	1.288 at 10°C
	1.292 at 4°C
At 21°C ...	1.280 Fully charged
	1.25 ¾ "
	1.22 ½ "
	1.19 ¼ "
	1.11 Flat
Dynamo belt ...	Tension by varying split pulley width maximum belt sag ½ inch

Generator/Dynamo

	From Engine No 056195	Up to Engine No 056195
Type: Sedan ...	DSV 90/12/16/3 S	R90–180/12–2500
Station wagon	D90/12/16/3 F	—
Nominal voltage ...	12 volts	12 volts
Maximum continuous output ...	16 amperes	13 amps
Maximum current output ...	22 amperes	—
Maximum continuous power ...	230 watts	180 watts
Maximum power ...	320 watts	—
Initial charging speed at 12 volts and 20°C ...	1710 to 1790 rpm	1300 to 1380 rpm
Continuous maximum output delivery speed at 20°C ...	2550 to 2800 rpm	2250 to 2400 rpm
Maximum output delivery speed at 20°C...	3050 to 3200 rpm	—
Maximum steady speed ...	9000 rpm	7500 rpm

	From Engine No 056195	Up to Engine No 056195
Rotation, drive end	Clockwise	Clockwise
Pole shoes	2	2
Field winding	Shunt	Shunt
Control box (separate)	FIAT GN 2/12/16	A/4–180/12
Drive ratio (new belt), engine-to-generator	1.74	1.74
Brush part number	4034356	879210

Bench testing data
—Testing generator as a motor (at 20°C):

	From Engine No 056195	Up to Engine No 056195
Feed voltage	12V	12V
Current draw	4.5 to 5.5 amps	4 to 4.5 amps
Speed	1500 ± 100 rpm	1050 ± 50 rpm

—Winding resistance test (at 20°C):

	From Engine No 056195	Up to Engine No 056195
Armature resistance	0.145 ± 0.01 ohms	0.31 ± 0.01 ohms
Field winding resistance	8.1 – 7.7 ohms	6.2 – 6.6 ohms

—Mechanical characteristics:

Load of spring on new brushes	1.3 to 1.6 lbs (0.600 to 0.720 kg)
Commutator maximum out-of-round	.00039 inch (0.01 mm)
Mica undercut depth	.0394 inch (1 mm)

Lubrication

Ball bearing, drive end head	FIAT MR 3 Grease or Castrol LM Grease	No change

Control box ... GN 1/12/16 and 2/12/16

Cutout relay ... From Engine No 056196
(Figures in brackets type A/4–180/12 of earlier cars)

Closing voltage	12.6 ± 0.2V
Reverse current: up to and not above	16 amps (10 amps)
Air gap (closed contacts)	.014 inch (0.35 mm)
Point gap	.015 inch ± .002 inch (0.45 ± 0.06 mm)

Voltage regulator

Battery	50 A/h
Half-load current	8 (6.5) ± 0.5 amps
Setting voltage after thermal stabilization in oven at 122° ± 5°F (50° ± 3°C) for 30 minutes, half-load on battery	14.2 ± 0.3 V
Feed voltage for thermal stabilization	15 V
Air gap	.039 to .04 inch (0.99 to 1.11 mm)

Current regulator

Regulated current on battery	16 ± 0.5 amps (13 ± 0.5 amps)
Voltage for regulated current inspection	13 V
Air gap	.039 to .044 inch (0.99 to 1.11 mm)

Regulating resistor

Types 1/12/16 and A/4–180/12	105 ± 3 ohms
Type 2/12/16	85 ± 5 ohms
Additional resistor in voltage regulator circuit on type 2/12/16	17 ± 1 ohms

Starter motor

Type	B 76–0.5/12 S
Voltage	12 V
Nominal power	0.5 kW
Rotation (pinion end)	Counter-clockwise
Pole shoes	4
Field winding	Series
Engagement	Pre-engaged: overrunning clutch
Pole shoes I.D	2.0697 to 2.0768 inch (52.57 to 52.75 mm)
Armature diameter	2.0394 to 2.0413 inch (51.80 to 51.85 mm)
Part number of brushes	805581

Bench test data
—Operation test at 68°F (20°C):

Current	130 amp
Torque developed	2 ± .14 lbf.ft (0.28 ± 0.02 kg.m)
Speed	2,250 ± 100 rpm
Voltage	10 volts

Chapter 8/Electrical system

—Stall torque at 68°F (20°C):
   Current ... 258 amp
   Voltage ... 7.7 ± 0.3 volts
   Torque developed ... 5.3 ± .36 ft.lbs (0.73 ± 0.05 kg.m)
—No-load test at 68°F (20°C):
   Current, not above ... 30 amp
   Voltage ... 12 volts
   Speed ... 8,500 ± 1,000 rpm
   Calculated resistance during stall torque test, at 68°F (20°C) ... 0.03 ± 0.001 ohms

Mechanical characteristics test
—Load of springs on new brushes ... 2.5 to 2.9 lbs (1.15 to 1.30 kg)
—Armature shaft axial play ... .0059 to .0256 inch (0.15 to 0.65 mm)
—Mica undercut depth, not above ... .04 inch (1 mm)
—Drive unit free wheel efficiency: static torque required to rotate pinion slowly, not above ... .35 lbf in. (0.4 kg.cm)

Lights
Headlamps: High beam ... 45 watts
          Low beam ... 40 "
Parking lights ... 5 "
Indicator flasher: Front/Rear ... 20 "
Indicator flasher: Side (tubular) ... 2.5 "
Brake warning ... 20 "
Number plate ... 5 "
Interior (on mirror frame) ... 3 "
Instruments ... 2.5 "
Warning lights on dash ... 2.5 "

Fuses
Six, of ... 8 amps
three different circuit allotments (see section 22)

1 General description

1 The electrical system is of 12 volts, and has negative earth.
2 The battery is in the luggage compartment at the front.
3 A DC generator is fitted. This also carries the engine cooling fan. As the latter is different to the station wagon this car has a different generator. The output is controlled by a full voltage and current regulator.
4 The starter is the pre-engaged type.
5 There is neither ammeter or voltmeter.
6 A full range of lights is fitted.

2 Battery inspection

1 Topping up the electrolyte is a monthly or 1,500 mile task.
2 Also monthly is the visual check for leaks, security, or corrosion.
3 The terminals will not corrode for years if properly protected. To do so, disconnect them, and take out the terminal bolt. Clean all very thoroughly. Smear thoroughly with vaseline, NOT grease. Reassemble, and wipe off the surplus vaseline.
4 The area near the battery may corrode. If it does treat it as described in section 6.3 and 4.

3 Electrolyte

1 The liquid in the battery is sulphuric acid.
2 It is highly corrosive to the car, you, and your clothes.
3 It should be kept topped up with pure water. The acid itself does not get used up; just the water bubbles off as the battery "gasses" when charging.
4 The "pure" water is normally got in the form of distilled water. It is far better to top it up with tap water than to let it get low for lack of the distilled.
5 Due to chemical reactions inside, the electrolyte specific gravity (sp gr) falls as the battery discharges. This can be measured with a hydrometer (see the specifications).
6 If the battery acid is spilled the lost electrolyte must be replaced with acid. This must be mixed to the correct sp gr. If diluted concentrated acid, pour the acid into the water. NEVER add water to acid: It will explode. First dilute with 1 part acid to 2½ parts water. Then continue till the required sp gr is achieved as appropriate for that temperature.
7 On old batteries the sp gr cannot be got back to that of a new one due to permanent chemical changes.

4 Charging

1 If the car is used frequently, on good journeys, and has a young battery, the generator will keep the battery fully charged.
2 Old batteries do not hold their charge.
3 Town journies do not give good charges due to idling at traffic halts.
4 If the battery is not kept fully charged its plates deteriorate faster than normal.
5 In winter the generator may well not keep the battery properly charged. It should therefore be recharged from an outside source at least once in the winter.
6 Charge at a rate not exceeding 3½ amps. When fully charged the battery gasses more freely, and the sp gr reaches a maximum. Continue to charge for about 2 hours after this.
7 Do not have a so called "boost charge", which takes only about 1 - 2 hours. This will shorten the life of the battery by a factor of many years.
8 An old battery can often be revived by "cycling" it once or twice. Let it discharge fully, slowly, by leaving the parking lamps on. Then charge it at only 1½ - 2 amps, till fully charged.
9 A reputable make of battery should last four to six years.
10 Without fitting an ammeter to the car it is impossible to tell how well the dynamo is charging the battery. This fitment is well worth while. As the battery reaches full charge its voltage rises quite sharply. The dynamo output will then cut back as the voltage regulator cuts in.

5 Battery leaks or corrosion

1 If the battery leaks, remove it immediately before the acid can do any more damage.
2 If the casing is cracked, take it to an expert to mend. It is not at all easy to get a repair to last. The corrosion from a leak is so severe that it is not a good risk to have an unreliable repair. Leaks in the jount round the top are more easily dealt with. A small one can be filled with a household sealant such as Sealstic, but for larger cracks melt some pitch.
3 Halt any corrosion in the car by washing out the front with plenty of water.
4 Once thoroughly dry, paint with a rust preventer such as zincote, followed by the normal undercoat and top coats of paint.

6 Dynamo - description

1 Once the work of removing the dynamo from the car is completed, it is easy to work on. Provided the routine inspection and preventative maintenance of cleaning the commutator and replacing the brushes is done as part of the 18,000 mile task it should last without problem the life of the car. Being a DC generator, it is referred to as a dynamo.
2 Due to its position on the cooling system trunking it is a long job to remove it.
3 The fan on the station wagon is at the rear end of the generator, so a different type is used.
4 The dynamo consists of an armature rotating in a magnet. The magnetism of this is provided by a field coil. Regulation of the output is done by controlling the voltage fed to the field coil.
5 To prevent the battery driving the generator as a motor, there is a cut out which disconnect the generator when it is not charging.

7 Generator fault finding

1 There can be two types of fault. The charging rate may become low, or the output may stop completely. A low output is difficult to detect unless an ammeter is fitted. The first symptoms are likely to be a flat battery, but a complete failure will be shown by the so called ignition warning light. This is a misnomer. Like the oil warning light it comes on when the ignition is switched on. It stays on till the generator is charging.
2 If, with the engine running, no charge comes from the dynamo, or the charge is very low, first check that the fan belt is in place and is not slipping. Then check that the leads from the control box to the dynamo are firmly attached and that one has not come loose from its terminal.
3 If wiring has recently been disconnected check that the leads have not been incorrectly fitted.
4 Make sure none of the electrical equipment such as the lights or radio, is on, and then take the leads off the dynamo terminals. Join the terminals together with a short length of wire.
5 Attach to the centre of this length of wire the positive clip of a 0-20 volts voltmeter and run the other clip to earth on the dynamo yoke. Start the engine and allow it to idle at approximately 1500° rpm. At this speed the dynamo should give a reading of about 15 volts on the voltmeter. This speed is a fast idle: Do not run the engine faster or the field winding may be overloaded.
6 If no reading is recorded then check the brushes and brush connections. If a very low reading of approximately 1 volt is observed then the field winding may be suspect.
7 If a reading of between 4 to 6 volts is recorded it is likely that the armature winding is at fault.
8 If a satisfactory reading is obtained, then the fault is either in the wiring or the control box. Reconnect the two wires onto the generator. They have different sized terminals, so cannot be muddled.
9 Take off the leads on terminals 51 and 67, the left two, on the regulator. These can be muddled, so make sure they are marked to prevent this.
10 Again join the two leads together and repeat the same test. If again it is successful, and there is full generator voltage, then those leads must be alright, and the fault is in the control box or the wiring beyond. Refit the leads and refer to section 12.
11 If the first test proved the generator faulty it must be removed for further testing. Remove it as described in section 8.
12 Check the resistance of the field windings. This is easier done on the bench, as an accurate ohmmeter or Wheatstone bridge must be used. See the specification. If the reading is very high there is an open circuit. If the reading is below the specification then a short circuit in indicated. If the field windings have gone, unless a visual inspection discloses an easily mended defect, a reconditioned unit will probably be needed. Replacement windings are unlikely to be available, quickly and are difficult to fit. But, it should be possible to get a serviceable body complete at a breakers, and put your armature into the other body.
13 If the field winding seems alright, then the fault must be deeper in. Dismantle the generator. See section 9.
14 The brushes could well be at fault. They should be checked now anyway and the commutator cleaned as in section 10. Assuming all is in order continue with the fault finding as follows.
15 Examine the commutator segments. If the armature has some burned out windings with short circuits there will be burns, and the windings may show signs of over heating.
16 Test the resistance of the windings by checking the resistance from segment to the segment at 180°. This is very low, so needs accurate measurement. If facilities are not available this is where you give up. But the point is to be able to prove that it is the armature at fault or not, so you know what to do. Again, a replacement armature may be difficult to get, and one from a breakers may have a bad commutator, so you may be forced to get a reconditioned one.
17 If you do decide to measure the resistance of the armature then one method is to feed a low voltage of 1 to 2 volts in and measure the current. $R = V/A$.

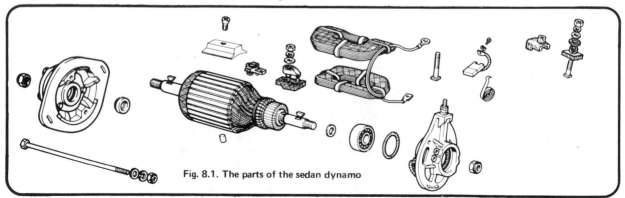

Fig. 8.1. The parts of the sedan dynamo

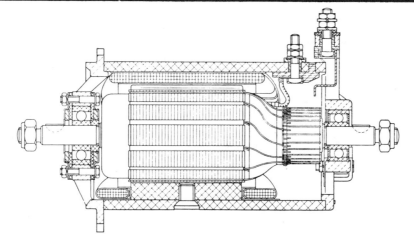

Fig. 8.2. The sedan dynamo: longitudinal section

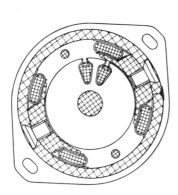

Fig. 8.3. Section through the windings

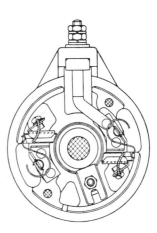

Fig. 8.4. Section through the brush end

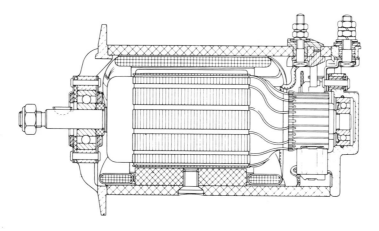

Fig. 8.5. Station wagon dynamo: it has different bearings because of the different fan

8 Dynamo - removal and replacement

1 The dynamo is mounted on the air cowling of the engine. Before it can be taken off the cowling, the fan must be removed from the far end of its shaft. If there are any other jobs to be done it might be better to remove the complete cowling, as described in Chapter 1.6. But assuming this is not so, proceed as follows:
2 Take off the fan belt and pulley by undoing the three little nuts clamping the two halves of the generator pulley together.
3 Disconnect the two leads from the dynamo, (photo).
4 Take off the air intake trunk from the cowling.
5 Reach down into the cowling and put a spanner on the nut holding the fan to the dynamo spindle, (photo).
6 Another spanner must be rigged up on the pulley end to hold the shaft. Put the pulley nuts back on to protect their threads. Undo the fan nut. Take care neither it nor its washers fall down into the trunking, (photo).
7 Undo the two nuts holding the dynamo to the cowling, (photo).
8 Slacken the bolt under the dynamo that tensions the strap holding it to the crankcase, till the pin can be pulled out of the top end of the strap, (photo).
9 The dynamo is now loose, but still with the fan on the end. Also loose is a shield underneath the dynamo. Retrieve this, noting how it is located by a pip pressed into it.
10 Pull the dynamo back to get the fan off the far end of the shaft. Make sure the woodruff key does not fall out. But take it out and put it in a safe place, (photo).
11 Refitting the dynamo is the reverse, but tighten the dynamo fixing to the cowling before fixing it into the strap on the crankcase.

9 Stripping the dynamo

1 Undo the nut holding the pulley flange to the shaft. (photo)
2 Pull off the flange, and take out the woodruff key, and put that somewhere safe; like in a jam jar, (photo).
3. Undo the two nuts holding the brush-end plate onto the body, (photo).
4 Carefully take off the brush-end end plate, so that the brushes come free gently, (photo).
5 Lift the armature out of the body. Take off the other end plate, and take out the two long bolts, (photo).

10 Dynamo - overhaul

1 It is assumed that the dynamo has not been stripped for some time. It will therefore need cleaning, relubrication of the bearings, and new brushes.
2 All this is part of its 18,000 mile maintenance task.
3 New brushes should be bought before starting work, as the need for them is almost a certainty. They should be fitted unless the existing ones are worn less than 1/8 in (3 mm).
4 Note the way the brushes and springs are fitted. Then remove them. Clean out the brush holders.
5 Do not immerse the dynamo end plates with the bearings in cleaning liquid, as dirt and liquid will not be fully removed. But wipe off the outside dirt. The design of the bearing houses varies, but unbolt or lift off one of the dirt shields: wipe away the old grease on the outside of the bearing, and push in some new grease. Check the bearings run smoothly and have no more than a trace of free movement.
6 The armature should be wiped clean. Take care the rag does not catch in and damage the windings.
7 Next check the condition of the commutator. If it is dirty and blackened, clean it with a rag just damped with petrol. If the commutator is in good condition the surface will be smooth and free from pits or burnt areas, and the insulated segments clearly defined.
8 Scrape the dirt out of the undercut gaps of insulator between the metal segments with a narrow screwdriver.
9 If, after the commutator has been cleaned, pits and burnt spots are still present, wrap a strip of fine glass paper round the commutator. Rub the patches off. Keep moving the paper along and turning the armature so that the rubbing is spread evenly all over. Finally repolish the commutator with metal polish such as Brasso. Then reclean the gaps.
10 In extreme cases of wear the commutator can be mounted in a lathe. With the lathe turning fast, take a very fine cut. Then polish with fine glass paper, followed by metal polish.
11 If the commutator is badly worn or has been skimmed the segments may be worn till level with the insulator in between. In this case the insulator must be undercut. This is done to a depth of 1/32 in (0.8 mm). The best tool is a hacksaw blade, if necessary ground down to make it thinner. The under cutting must take the full width of the insulator away, right out to the metal segments on each side.
12 Again clean all thoroughly when finished, and ensure no rough edges are left. Any roughness will cause bad brush wear. In all this work it must also be remembered that the commutator is not very strong.

11 Reassembling the dynamo

1 Put onto the armature first the brush end plate, with the brushes in place, pushing them back with clean fingers against the springs to get them over the commutator, (photo).
2 Now lower down the body of the dynamo over the armature. Slide the nylon tongue that is under the small terminal on the body into the post for the larger on the end plate, and also locate the dowel on the body.
3 Now fit the other end plate. Again locate the dowel: These dowels are very small peenings on the body, locating with grooves in the end plate.
4 Guide the ends of the long bolts through the holes in the end plate. Do this with the dynamo upright so that the bolts will hang vertically and find their way through the holes.
5 Fit the flat washers and the self locking nuts to the bolts.

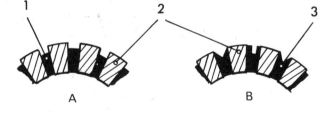

Fig. 8.6. Undercutting a commutator: A Correct — B Wrong, the cuts being too narrow.

1 Insulator
2 Segments
3 Insulator

8.2. Disconnect the leads and take off the belt and both halves of the pulley

8.5. Take off the air intake trunk. Reach round behind the engine and undo the fan nut

8.6. Hold the dynamo shaft whilst undoing the fan nut

8.7. Undo the nuts: We took the cowling off to take the photo

8.8. Slacken the strap by the bolt underneath and take the pin out of the top end

8.10. Pull the dynamo to get the fan off

9.1. Undo the nut

9.2. Take off the pulley and remove the woodruff key

9.3. Undo the nuts

9.4. Take off the brush-end plate

9.5. Lift the body off the armature: and then take off the other end plate

11.1. When it comes to reassembly fit the brush-end to the armature first; so the brushes can be slid easily over the commutator

12 Control box - function and faults

1 The control box has three relays. One is the cut out, another limits the dynamo voltage generated, and the third its current.
2 The cut out disconnects the dynamo from the battery when it is no longer charging, otherwise it could run off the battery as a motor.
3 The two regulator relays by their combination of voltage and current control regulate the output to suit the electrical load such as lights that might be switched on, and to suit the state of charge of the battery.
4 The current control is set to the maximum safe limit for the dynamo. The voltage regulator is set to a potential that will limit the charge given a full battery to a mere trickle.
5 If the control box has a complete failure the ignition warning light will come on. If there is a partial failure, unless an ammeter is fitted, there will be no warning. Undercharging may become apparent as a flat battery. Minor overcharging will give the need for frequent topping up of the battery. Gross overcharging may blow light bulbs, and perhaps result in a smell of burning, from the overloaded dynamo.
6 Major defects are likely to be the burning of the points on the relays, so that they never make contact: So no regulation takes place; or the wiring may burn out.
7 Minor defects occur due to wear and general ageing altering the voltage/current at which the cutout or regulators work.
8 A car-type ammeter will show these aberrations, but for fault finding more accurate instruments are needed. Unless you have some experience of such things, and of the instruments, it is suggested you do not tamper with the control box. If done incorrectly a new control bow and a new dynamo may be needed.
9 In any work on the control box it is important that the leads are not fitted to wrong terminals or the unit will be ruined.

Fig. 8.7. The control box showing the terminal numbering

13 Cut out checks

1 The cut out is the relay on the right as you look at them: the only one of the three with points open when the engine is switched off.
2 The dynamo will have been proved by earlier fault finding.
3 Take off the control box cover.
4 Check the voltage at the terminal from the dynamo, number 51, and the output from the cutout, number 30. This will prove that the defect is at the cutout.
5 Assuming there is 12/15 volts at the terminal 51, but none at 30, check the operation of the relay of the cutout. With the engine running fast enough to charge, try pushing it with a finger.
6 If the relay does not hold down there is a fault in the wiring.
7 If it stays down but there is still no voltage at the output terminal 30 the points appear to need cleaning. If the push made it work it may need resetting.
8 To reset the cut out wire a voltmeter from terminal 51 (the dynamo connection) to earth. Start up: Warm up for 15 minutes. Increase engine speed gradually, watching both the voltmeter and the cut out. Note the voltage at which the cut out closes; there will be a little kick of the voltmeter. It should close at 12.6 volts. This voltage should be set after the cut out is warmed up by about 15 minutes running.
9 Adjust by bending the arm on which the spring of the contacts rests, increasing the spring tension to raise the operating voltage.
10 On slowing down the cut out should "drop-off" that is cut out when the dynamo stops charging. The reverse current should never be high: The official maximum is 16 amps, which is high as such things go. To improve drop-off bend the fixed contacts so that the moving one cannot be drawn so close to the armature. A car ammeter will show this negative current before drop off.

Fig. 8.8. As you look at it, voltage regulator on the left (above terminal 51), current control in the centre, and cut-out right (above terminal 30).

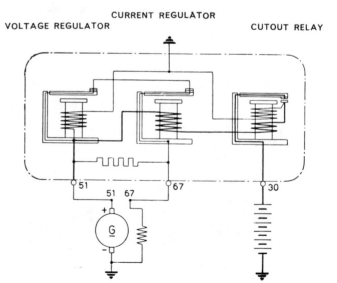

Fig. 8.9. The wiring diagram of the regulator

Chapter 8/Electrical system

14 Voltage regulator checks

1 The circuit needed is shown in Fig. 8.12.
2 This can easily be achieved by sliding a piece of paper in between the cutout points, and connecting the voltmeter from the dynamo connection terminal 51 on the left of the control box and to earth. This must be an instrument accurate to 0.3 volt.
3 Start up, and run the engine fast: at 3,000 rpm. The regulator should limit the voltage to 15.5 volts. It is important to take the reading quickly to avoid temperature effects.
4 The reading should be steady. Fluctuations imply the contact points need cleaning. This should be done with fine glass paper, and all dust removed.
5 If adjustment is needed bend the arm onto which the spring blade rests, increasing the spring tension to raise the regulated voltage. The voltage regulator is the one farthest from the cut-out.
6 The FIAT setting procedure involves the removal of the regulator and its preheating in an oven. The setting figure of 15.5 volts quoted is a compromise. It is valid at an ambient temperature of $10^{o}C$. At $20^{o}C$ set to 15 volts. These voltages are the maximum. Do not exceed this. If the car is used extensively on long journeys a voltage lower by 0.5 volts should be used.

15 Current regulator checks

1 The current control is set using an ammeter accurate to 0.5 amp. The circuit is as shown in the figure, the ammeter being wired between the control box output terminal number 30 and the leads that are normally connected to it.
2 Wedge cardboard between the voltage regulator armature (the one furthest from the cutout) and its arm to hold the contacts closed, so no voltage regulation can take place.
3 Start up the engine.
4 Turn on the headlamps to load the dynamo.
5 Speed up the engine to a fast speed of about 3,000 rpm.
6 The maximum current should be 16 amps (13 amps for early cars - see the specifications).
7 The current should be steady. Fluctuations imply the contacts need cleaning with fine glass paper, and all dust removed.
8 If adjustment is needed bend the arm onto which the spring blade rests, increasing the spring tension to raise the controlled current. The current regulator is the centre one.

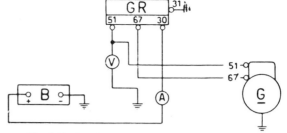

Fig. 8.10. The wiring diagram for checking the cut-out

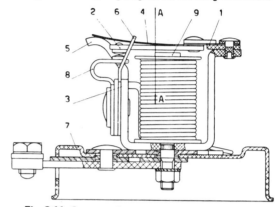

Fig. 8.11. Cut-out relay

1 Spring hinge
2 Moving contact
3 Body
4 Bi-metal spring
5 Adjustment arm
6 Stop
7 Base
8 Fixed contact
9 Armature core

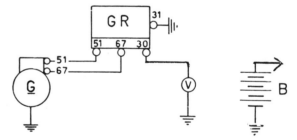

Fig. 8.12. Wiring diagram for setting the voltage regulator

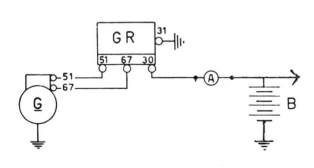

Fig. 8.13. Wiring diagram for setting the current regulator

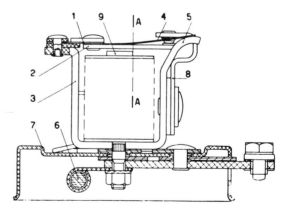

Fig. 8.14. Regulator relays

1 Hinge
2 Moving contact
3 Body
4 Spring
5 Adjustment arm
6 Regulating resistance
7 Base
8 Fixed contact
9 Armature core

16 Starter motor - description

1 The starter is of the pre-engaged type.

2 When the cable is pulled by the lever beside the driver an arm on the top of the starter does two things. A fork at the bottom of the actuating lever on the starter body slides the drive gear on the starter shaft towards the gear ring on the outside of the flywheel. If the teeth happen to be lined up it goes into mesh. If not lined up, the starter's gear slides on its mounting against a spring, which will push it into mesh as soon as the starter begins to turn and lines up the teeth. Once the lever has got near the end of its travel the gears are ready for the motor to turn. In the last bit of its movement the lever pushes on a switch and connects the motor.

3 Once the engine fires, the starter motor could run too fast as the engine picks up, so there is a freewheel in the drive.

4 More sophistocated starters have a solenoid so that the mechanism is operated on a definite basis. It is possible on this one on the FIAT 500 to operate the control half-heartedly, so that the switch is not firmly closed. This will harm the switch, and may give trouble.

17 Removing and refitting the starter motor

1 Disconnect the positive battery lead.

2 Reaching round the engine it is possible to get at the starter on the sedan. It is much easier on the station wagon. It may be best on the sedan to have a good look with a mirror first, and then do everything by feel.

3 Disconnect the actuating cable from the lever on the starter top, by removing the pin.

4 Unbolt the electric cable (13 mm spanner) from the starter.

5 Undo the nuts (2 on early cars, 3 on later ones) holding the starter to the engine.

6 Lift out the starter.

7 Refitting the starter is the reverse.

8 On early cars the lever has two holes. Later cars only have the hole in the end, and this is the one to use on the early ones. The fitting on the end of the cable has three holes to allow adjustment. Put the lever's pin through the tightest hole possible without moving the starter lever from the rest position.

18 Stripping the starter motor

1 Remove the two Phillips headed screws that hold the starter switch to the top of the body, and take off the switch. (photos)
2 Undo the screw clamping the shield over the apertures for the brushes and remove the shield, (photo).
3 Undo the two nuts on the long bolts that clamp the whole starter together, at the brush end, (photo).
4 From the other end pull off the starter drive, with its engaging mechanism, (photo).
5 At the windows for the brushes, undo the terminal securing the field coils in the starter body to the brush end plate. Then take off the end plate, (photos).
6 To dismantle the drive mechanism, first remove the dirt shield. On early models this is a three piece metal one held by a screw to the lever. Later ones have a rubber boot held by the pin of the actuating lever, (photo).
7 Remove the lever's pin, (photo).
8 Push the drive pinion as far as it will go into the housing. Disconnect the forks of the engaging arm from the groove in the drive pinion, and then bring out the engaging lever fork-end first. The components of the drive can now be lifted out, (photo).
9 Unscrew the switch from the top of the starter body.

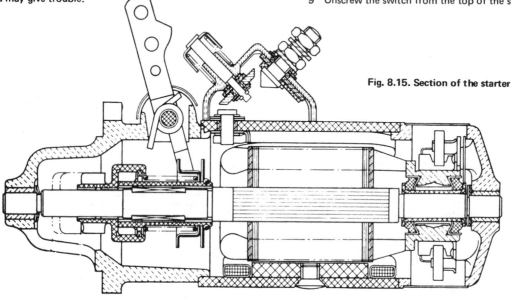

Fig. 8.15. Section of the starter

19 Checking the starter motor

1 Check the brushes for wear: It is probably best to renew them unless they are worn less than 1/8 in (3 mm).
2 Clean the commutator and clear out the segment gaps. Scrape the dirt out with a sharp screwdriver. If the commutator is in bad condition it can be cleaned up as was described for the dynamo in section 10.
3 Check the bearings for wear. Clean them and relubricate them.
4 Clean all the components of the drive, and check them for wear. Lubricate with a molybdenum - disulphide grease.
5 If any of the major components are badly worn, it may be best to get a replacement unit.
6 The bearings are "self lubricating". Rewet them with engine oil, allowing it time to soak in before reassembly.
7 Check the switch contacts for burning. If the starter control lever has been pulled gently too often the switch will have been burned. With gentle operation the two contacts are not firmly pressed against each other, so sparking takes place. This can lead either to the starter failing to work, or worse, not stopping when the control is released. Should the latter happen, switch off the ignition, and take off a battery lead quickly.

18.1a. Undo the screws

18.1b. and remove the switch

18.2. Then the shield

18.3. Undo the two nuts

18.4. From the other end take off the drive

18.5a. Disconnect the field windings

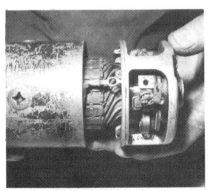
18.5b. The body,

18.5c. and brush-end plate can now be taken off

18.5d. Don't miss the washer

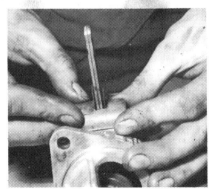
18.6. Take off the dirt shield (its rubber on later cars)

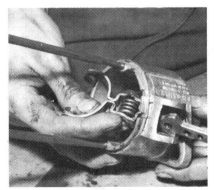
18.7. Take out the pin and wriggle out the lever

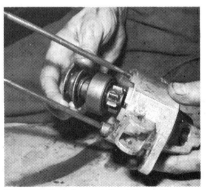
18.8. Finally bring out the drive unit

8 Fault finding is different from that on the dynamo. The starter field is in series with the armature. So the motor should be dismantled and the resistance of both checked individually. If the field has some turns short-circuited, the reduced magnetic effect will give a loss of power.
9 These starters do have undercut commutators.

20 Reassembling the starter

1 Prepare the drive unit. Drop the drive pinion down the end of the housing.
2 Thread the lever into place. Note that it must be that way round which will bring the lug on the lever arm towards the main body of the starter, for pressing down the switch. Having got it into place push the spring into its place in the double part of the lever arm, and thread in the retaining pin.
3 The arms of the spring have got to be hooked on so that they will hold the lever in the disengaged, and off, position.
4 Insert the pin through the lever and the split pin through that. Put on the metal dirt cover: It wants to start down the lever at 90° to its correct position, and then be turned to fit over the boss on the top of the starter. Fit the two side pieces at the bottom of the lever to complete the dirt shield, and hold them with their screw. Starters with the rubber boot should have a new one fitted, round the actuating lever pin.
5 Fit the washer to the shaft, at the commutator. Put the brush end plate over the commutator, holding the brushes up against their springs with clean fingers as it slips over.
6 Fit the body over the armature, sliding it down towards the brush-end plate with the field winding terminal lined up with its fixing in the end plate. Put the screw through the connecting tag for the brush and the terminal for the field winding, and screw it into the post on the end plate.
7 Fit the drive end. Note that the two long bolts that will hold the whole starter together are covered with insulating sleeves. As the drive unit is slid on it will be necessary to turn the drive pinion to line up the splines.
8 Put the flat washers on the ends of the long bolts, and fit the self locking nuts and tighten them.
9 Slide the dirt shield over the brush end to cover the windows. Fix it so that the slit at the end is not by one of the windows.
10 Refit the switch to the top of the starter.

21 Fuses

1 All models have six fuses. But the circuits wired to each fuse have changed during the models production run.
2 Fuses normally blow very infrequently. If a bulb fails it may blow the fuse as it goes. Fuse blowing is usually easily recognised because of the other electric failures coinciding. The only exception are those fuses working only one item.
3 If a fuse consistently blows, it is indicating a short in that circuit, usually intermittent. DO NOT be tempted to replace the fuse by thicker wire. The result will be that as the short gets worse, the wiring of the car will overheat, causing widespread harm, and even burning the whole car.
4 Whilst searching for the cause of blowing fuses it is economical to use household fuse wire laid in the clips and held by the burned out fuse cartridge.
5 The cause of the trouble is likely to be a frayed wire, chafing where it passes through a hole in sheet metal.
6 "New 500" (and 500 Sport) Fuses:

FUSES	PROTECTED CIRCUITS
1 - Fuse No. 30/2	Right headlamp high beam - Left front parking light with indicator - Right rear parking light.
2 - Fuse No. 30/3	Left headlamp high beam with indicator - Right front parking light - Left rear parking light - Number plate light.
3 - Fuse No. 56/b1	Left headlamp low beam.
4 - Fuse No. 56/b2	Right headlamp low beam.
5 - Fuse No. 15/54	Winking direction indicators with pilot light - Instrument cluster light - Stop light.
6 - Fuse No. 30	Horn - Windshield wiper - Rear view mirror light.

Unprotected circuits:
Generator charge and relevant indicator.
Ignition.
Starting.
Low oil pressure indicator.
Fuel reserve supply indicator.

7 500D and station wagon fuses:

FUSES	PROTECTED CIRCUITS
1 - No. 30/2	Right headlamp high beam. Front left parking lamp. Rear right parking light. Front parking lamps indicator. Number plate lamp.
2 - No. 30/3	Left headlamp high beam - Front right parking lamp - Rear left parking light. High beam indicator.
3 - No. 56/b1	Left headlamp low beam.
4 - No. 56/b2	Right headlamp low beam.
5 - No. 15/54	Direction indicators and pilot light. Panel light. Stop lights. Windshield wiper.
6 - No. 30	Horn. Lamp in rear view mirror. Rear inner lamp (Station wagon only).

The unprotected circuits as in paragraph 6 above.

8 500 L sedan fuses:

FUSE	PROTECTED CIRCUITS
1 - A	— Horn — Light in rear view mirror.
2 - B	— Direction indicators and repeater. — Low oil pressure warning light. — Fuel gauge and low fuel warning light. — Windshield wiper. — Rear stop lights.
3 - C	— Right side low beam.
4 - D	— Left side low beam.
5 - E	— Left side high beam and indicator. — Front right parking light. — Rear left parking light. — Instrument cluster light.
6 - F	— Right side high beam. — Front left parking light and indicator. — Rear right parking light. — Number plate light.

Unprotected circuits are the charging and its warning light, the ignition, and the starter motor.

9 The fuse box is mounted in the luggage locker, at the left front of the car, beside the fuel tank.
10 Fuse number one is that one at the rear end, nearest the windscreen.

22 Windscreen wiper

1 Proving a fault is in the motor and not the wiring is made difficult by the otherwise excellent plug connecting the leads to the motor. Probes will be needed to reach in with a test voltmeter or bulb into the plug.
2 The wiring circuit for the motor and its self parking arrangement is given in Fig.8.11.
3 For parking an additional field winding for the motor is connected. On a DC electric motor increasing the field strength gives stronger torque, but slows the motor. In this case this will give the strength to the motor to get the blades over mud or snow to the parked position, and also slow it so that it is unlikely to overshoot the park position.
4 Once a defect in the motor is suspected, it must be removed with the blade mechanism, all on the mounting plate, (photo).
5 Disconnect the plug for the leads, (photo).
6 Pull off the wiper arms, (photo).
7 Take off the fixtures on the ends of the arm spindles so that these can be withdrawn through the bushes in the body below the windscreen, (photo).
8 Undo the two bolts holding the motor mounting bracket to the bulkhead, (photo).
9 Remove the whole assembly, (photo).
10 When refitting, grease the arm spindles in their bushes, and all the other links for the connecting levers. Operate the motor before fitting the blade arms. Once the motor has put the spindles in the parked position, the arms can be fitted in their correct orientation.
11 If the operation of the wiper becomes very sluggish due to the need for lubrication of the interconnecting mechanism, or in heavy snow, there is risk of burning out the motor. Always switch off the wiper if it stalls, and if it cannot get back to the parked position, unplug the motor.

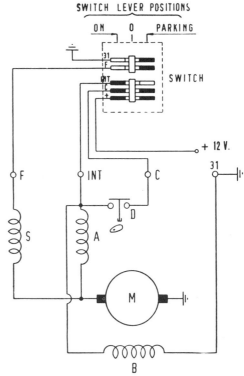

Fig. 8.16. Windscreen wiper wiring diagram
A Series winding. B Shunt winding. D Switch. M Motor
S Additional parking field winding. F INT
C Terminal markings

22.4. The wiper motor is just behind the petrol tank

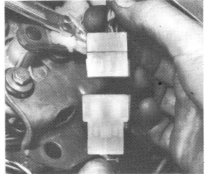

22.5. Unplug the leads

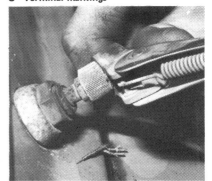

22.6. Pull off both wiper arms

22.7. Take off the fixtures for the arms. You may also have to take out the bush in the body too if it is old and the spindle won't come free

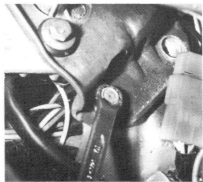

22.8. Undo the two bolts holding the mounting to the body

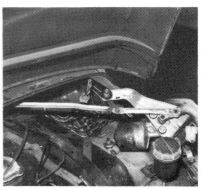

22.9. Lift out the whole assembly

23 Horn

1 If the horn fails to work first check whether it is the horn button or the actual horn.
2 The horn button ring is a push fit in the steering wheel. If the problem is here it may be burned points due to sparking as the button makes contact.
3 The circuit from the button is to a contact by the indicator stalk on the steering column. This contact must be good. A failure here is normally given away by the horn working fitfully when the steering wheel is wiggled. The contact should be coated with vaseline.
4 The horn button is the earth return, and is the black/yellow wire. Find which it is at the horn. Check the other lead is live with a voltmeter or test lamp. Then earth the earth contact with an odd length of wire, to prove the horn itself is working.
5 The more recent the horn the less easy it is to mend. Early ones have two adjusters. One is in the centre of the diaphragm, a centre screw with outside it a locknut. On the other side of the body, off centre, is another adjuster. Before moving an adjuster, mark its position. Work the horn button, and listen for a faint click as the horn tries to work. Try the effect of altering the adjusters a small amount. Moving them may disturb some dirt and get the horn going again. Adjust the screws for the strongest note. Then daub them with paint to seal out damp and to lock them in position. If this fails, some horns can be stripped to mend them. Ones rivetted together can but be replaced.

24 Lighting

1 Many cars are to be seen driving with some lamps inoperative.
2 As bulbs blow without warning and at inconvenient times, always carry spare ones on the car.
3 It is possible to check bulbs on the move by reflections in the body panels of vans or in shop windows. But the surest way to check the tail lamps is to walk round and look at them. The small lights at the front are parking lights only, so these will be checked on leaving the car. They are of no use when driving, being too dim to be used in fog in daylight or under any dark conditions.
4 If an indicator flashing light bulb fails, it will give a different flashing speed to the winkers.
5 Except for the headlamp low beams which have individual fuses, the failure of one lamp only is usually the bulb. Only very occasionally is it a wiring failure, as these failures are usually a short circuit rather than an open one. But failure of the earth contact at the light afflicts cars as they get old.
6 Remember in checking the brake lights the ignition must be on. These bulbs are a common failure; they can be checked by reversing close to a wall.

25 Headlight - aiming

1 The aim of the headlights is set by the mounting screws on the back.
2 When making any changes count how many part turns they are moved. Early cars have three screws, the top one giving vertical adjustment, and the bottom two horizontal. If making a horizontal correction one screw must be screwed one way and the other the opposite, so that no vertical change is made. Later cars have two screws, the top one again being for vertical correction.
3 In the diagram is the layout for using a wall to set the lights on low beam. See Fig. 8.13.b.
4 The setting can also be seen well in mist.
5 If on coming traffic flashes at you it is a sure sign your lights are badly set. If you cannot get deserted conditions in which to set them yourself then it is best to get a garage to do it with a beam setting gauge.

26 Changing a headlight unit

1 On early cars with the three headlight aiming screws, these also secure the light unit to the car. To get the unit out, push it in against the springs around the adjusters, and turn it anti-clockwise as far as it will go, to get the adjuster/mounting screws round to the larger holes. If the light unit is stiff to turn all the adjusters may be a bit tight. If so unscrew them all an equal amount.
2 Later cars have the units held in the mudguard by tongues sticking outwards from the light unit mounting. These must be pushes inwards to disengage them from the mudguard.

27 Fitting accessories

1 The FIAT 500 has a proper electric circuit so can take all normal accessories.
2 It is suggested extra driving lights would be inappropriate for such a car, but all other things will be just as welcome, particularly an ammeter.
3 To connect an ammeter remove the brown lead from terminal 30 on the control box. This is the one from the battery via the starter. Connect this wire to one side of the ammeter, and connect a new lead back from the ammeter to the terminal number 30. If the ammeter shows a charge when the lights are turned on change the wires over on the ammeter.
4 All accessories must be well wired with firm joins. If you have no facilities for soldering a good idea is to buy little screwed terminal blocks from any household electrical shop. A break can then be made in a convenient run of the car's own wire, an the terminal inserted to act as a junction point. Wires passing through holes in metal must have a grommet to protect them. The additional wiring must pick up at a point protected by a fuse.
5 Such things as reversing lamps should be wired up from the lighting circuit, so they cannot be left on by mistake.

28 Instruments

1 Access to these can be gained from the luggage compartment.
2 Additional instruments can be mounted in the dash.

29 Fault finding principles

1 The tracing of an electrical fault follows the usual principle of the methodical check along the system.
2 First check for foolish errors, such as the wrong switch turned on, or for such things as an over-riding control like the ignition may be needed. Also, have other things failed; is it a fuse?
3 If it is a light it is now best to assume it is the bulb and change that.
4 The next thing is to have a quick glance for obvious faults such as a loose lead. On an old car it is regretted that a good kick, or at least a judicious blow, is sometimes needed next: If the earth return of a component is poor this will strike a contact. If it does, strip the component and derust it.
5 The proper systematic tracing of a more elusive fault requires a voltmeter or a test lamp. The latter is a small 12 volt bulb with two wires fitted to it; either using a bulb holder, or soldered direct. To test a circuit one wire is put to earth, and the other used to test the live side of a component. To test the earth side, the test bulb is wired to terminal 30 on the control box. All this is helped if the tester has crocodile clips on the wire.
6 Trace methodically from a component back along the circuit till the correct result is got. Inaccessible circuits can be bridged with a temporary wire, and this tried to see if it gives a cure.
7 On younger cars the wiring colour will still be visible. Where this has faded fault finding is more difficult, and substitute wiring has to be put in at an earlier stage. If it is decided the wiring is at fault the new wire must be securely fixed so that it will not chafe on anything.

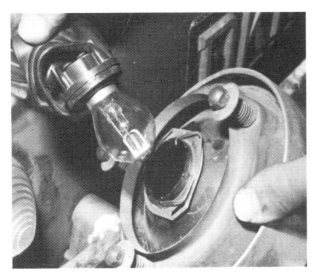

Fig. 8.17. Headlamp bulb (We have taken the light unit out just for the photo).

Fig. 8.18. Flasher side repeater

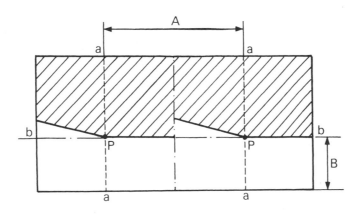

Fig. 8.20. Headlamp aiming wall marks. Distance apart, A = 83 cm. B = Height of spot. Distance of light centre of the car above ground less 4 cm for sedans and 7.5 cm for station wagons

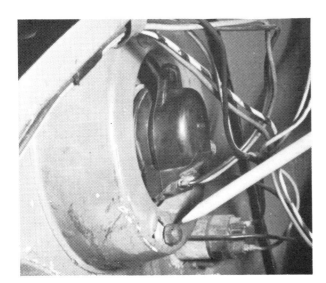

Fig. 8.19. The headlamp adjusters

Fig. 8.21. Wiring to instruments and warning lights can be reached from the boot.

30 Fault finding table

Some of the electrical problems were included in the engine fault finding chart in Chapter 1. This is a table of other troubles that can occur in the electrical system.

Cause	Possible trouble	Remedy
STARTER MOTOR FAILS TO TURN ENGINE		
No electricity at starter motor	Battery discharged	Charge battery.
	Battery defective internally	Fit new battery.
	Battery terminal leads loose or earth lead not securely attached to body	Check and tighten leads.
	Loose or broken connections in starter motor circuit	Check all connections and tighten any that are loose.
	Starter motor switch or solenoid faulty	Test and replace faulty components with new.
Electricity at starter motor: faulty motor	Starter motor pinion jammed in mesh with flywheel gear ring	Disengage pinion by turning squared end of armature shaft.
	Starter brushes badly worn, sticking, or brush wires loose	Examine brushes, replace as necessary, and tighten down brush wires.
	Commutator dirty, worn, or burnt	Clean commutator, recut if badly burnt.
	Starter motor armature faulty	Overhaul starter motor, fit new armature.
	Field coils earthed	Overhaul starter motor.
STARTER MOTOR TURNS ENGINE VERY SLOWLY		
Electrical defects	Battery in discharged condition	Charge battery.
	Starter brushes badly worn, sticking, or brush wires loose	Examine brushes, replace as necessary, and tighten down brush wires.
STARTER MOTOR NOISY OR EXCESSIVELY ROUGH ENGAGEMENT		
Lack of attention or mechanical damage	Pinion or flywheel gear teeth broken or worn	Fit new flywheel, or new pinion to starter motor drive.
	Starter drive broken	Dismantle and fit new parts.
	Starter motor retaining bolts loose	Tighten starter motor securing bolts. Fit new spring washer if necessary.
BATTERY WILL NOT HOLD CHARGE FOR MORE THAN A FEW DAYS		
Wear or damage	Battery defective internally	Remove and fit new battery.
	Electrolyte level too low or electrolyte too weak due to leakage	Top up electrolyte level to just above plates.
	Plate separators no longer fully effective	Remove and fit new battery.
	Battery plates severely sulphated	Remove and fit new battery.
Insufficient current flow to keep battery charged	Fan/dynamo belt slipping	Check belt for wear, replace if necessary, and tighten.
	Dynamo not charging properly	Remove and overhaul dynamo.
	Short in lighting circuit causing continual battery drain	Trace and rectify.
	Regulator unit not working correctly	Check setting, clean, and replace if defective.
IGNITION LIGHT FAILS TO GO OUT, BATTERY RUNS FLAT IN A FEW DAYS		
Dynamo not charging	Fan belt loose and slipping, or broken	Check, replace, and tighten as necessary.
	Brushes worn, sticking, broken, or dirty	Examine, clean, or replace brushes as necessary.
	Brush springs weak or broken	Examine and test. Replace as necessary.
	Commutator dirty, greasy, worn, or burnt	Clean commutator and undercut segment separators.
	Commutator bars shorting	Undercut segment separations.
	Dynamo bearings badly worn	Overhaul dynamo, fit new bearings.
	Dynamo field coils burnt, open, or shorted, or armature the same	Remove and fit rebuilt dynamo.
Regulator or cut-out fails to work correctly	Regulator incorrectly set	Adjust regulator correctly.
	Cut-out incorrectly set	Adjust cut-out correctly.
	Open circuit in wiring of cut-out and regulator unit	Remove, examine, and renew as necessary.
	Contacts dirty	Clean, re-adjust.

Chapter 8/Electrical system

Failure of individual electrical equipment to function correctly is dealt with alphabetically, item by item, under the headings listed below:

LOW FUEL WARNING - FUEL GAUGE (500L ONLY)

Fuel gauge gives no reading	Fuel tank empty!	Fill fuel tank.
	Electric cable between tank sender unit and gauge earthed or loose	Check cable for earthing and joints for tightness.
	Fuel gauge case not earthed	Ensure case is well earthed.
	Fuel gauge supply cable interrupted	Check and replace cable if necessary.
	Fuel gauge unit broken	Replace fuel gauge.
Fuel gauge registers full all the time	Electric cable between tank unit and gauge broken or disconnected	Check over cable and repair as necessary.

HORN

Horn operates all the time	Horn push either earthed or stuck down	Disconnect battery cable. Check and rectify source of trouble.
	Horn cable to horn push earthed	Disconnect battery cable. Check and rectify source of trouble.
Horn fails to operate	Blown fuse	Check and renew if broken. Ascertain cause.
	Cable or cable connection loose, broken or disconnected	Check all connections for tightness and cables for breaks.
	Horn has an internal fault	Remove and overhaul horn.
Horn emits intermittent or unsatisfactory noise	Cable connections loose	Check and tighten all connections.
	Horn incorrectly adjusted	Adjust horn until best note obtained.

LIGHTS

Lights do not come on	If engine not running, battery discharged	Charge battery.
	Light bulb filament burnt out or bulbs broken	Test bulbs in live bulb holder.
	Wire connections loose, disconnected or broken	Check all connections for tightness and wire cable for breaks.
	Light switch shorting or otherwise faulty	By-pass light switch to ascertain if fault is in switch and fit new switch as appropriate.
Lights come on but fade out	If engine not running battery discharged	Push-start car, and charge battery.
Lights give very poor illumination	Lamp glasses dirty	Clean glasses.
	Reflector tarnished or dirty	Fit new reflectors.
	Lamps badly out of adjustment	Adjust lamps correctly.
	Incorrect bulb with too low wattage fitted	Remove bulb and replace with correct grade.
	Existing bulbs old and badly discoloured	Renew bulb units.
	Electrical wiring too thin not allowing full current to pass	Re-wire lighting system.
Lights work erratically - flashing on and off, especially over bumps	Battery terminals or earth connection loose	Tighten battery terminals and earth connection.
	Lights not earthing properly	Examine and de-rust contact.
	Contacts in light switch faulty	By-pass light switch to ascertain if fault is in switch and fit new switch as appropriate.

WIPERS

Wiper motor fails to work	Blown fuse	Check and replace fuse if necessary.
	Wire connections loose, disconnected, or broken	Check wiper wiring. Tighten loose connections.
	Brushes badly worn	Remove and fit new brushes.
	Armature worn or faulty	If electricity at wiper motor remove and overhaul and fit replacement armature.
	Field coils faulty	Purchase reconditioned wiper motor.
Wiper motor works very slow and takes excessive current	Commutator burnt	Clean commutator thoroughly.
	Drive to wipers bent or unlubricated	Examine drive and straighten out severe curvature. Lubricate.
	Spindle binding or damaged	Remove, overhaul, or fit replacement.
	Armature bearings dry or unaligned	Remove, overhaul, or fit replacement.
	Armature badly worn or faulty	Remove, overhaul, or fit replacement.
Wiper motor works slowly and takes little current	Brushes badly worn	Remove and fit new brushes.
	Commutator dirty	Clean commutator thoroughly.
	Armature badly worn or faulty	Remove and overhaul armature or fit replacement.
Wiper motor works but wiper blades remain static	Driving gear damaged	Examine, and if faulty, replace.
	Wiper motor gearbox parts badly worn	Overhaul or fit new gearbox.

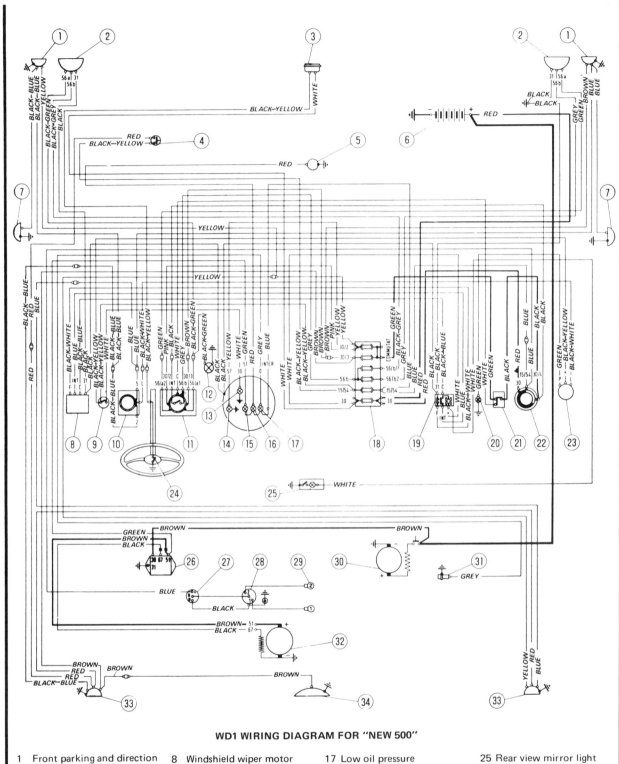

WD1 WIRING DIAGRAM FOR "NEW 500"

1. Front parking and direction indicator lamps
2. Headlamps (high and low beams)
3. Horn
4. Stop lights pressure - operated switch
5. Fuel reserve supply indicator sending unit
6. Battery
7. Side direction indicator lamps
8. Windshield wiper motor
9. Panel light switch
10. Directional signal lever switch
11. External lighting change-over lever
12. High beam indicator
13. Panel light
14. Parking lights indicator
15. Generator charge indicator
16. Fuel reserve supply indicator
17. Low oil pressure indicator
18. Fuses
19. Windshield wiper 3 - position switch
20. Direction indicator pilot light
21. External light switch
22. Ignition lock switch
23. Flasher unit, direction indicators
24. Horn button
25. Rear view mirror light
26. Generator regulator
27. Ignition coil
28. Ignition distributor
29. Spark plugs
30. Starter motor
31. Low oil pressure indicator sending unit
32. Generator
33. Rear parking, stop and direction indicator lamps
34. Number plate lamp

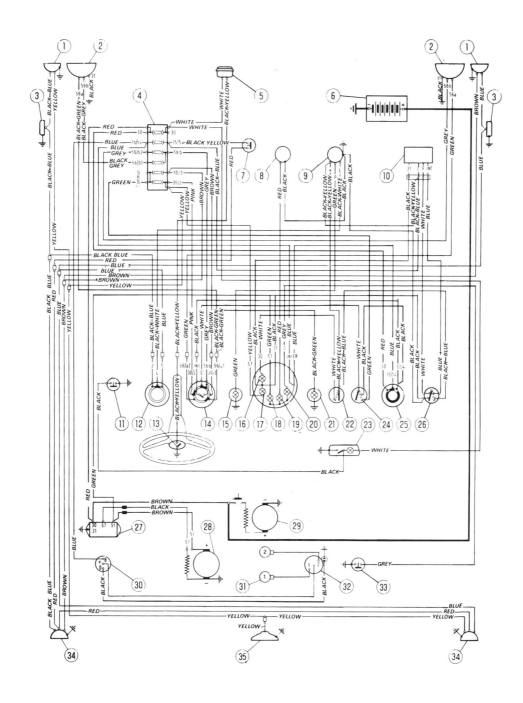

WD2 110D AND 110F SEDAN WIRING DIAGRAMS. THE 120 (STATION WAGON) HAS AN ADDITIONAL INTERIOR LAMP WIRED FROM FUSE 30

1. Front parking and directional signal lights
2. High and low beam headlights
3. Side directional signal lights
4. Fuses
5. Horn
6. Battery
7. Stop light pressure - operated switch
8. Fuel reserve supply indicator sending unit
9. Directional signal flasher unit
10. Wiper motor
11. Courtesy light jam switch on driver's side door pillar
12. Directional signal switch
13. Horn button
14. Change - over switch for lighting and flashers
15. Directional signal indicator
16. Instrument light
17. Parking light indicator
18. No-charge indicator
19. Fuel reserve supply indicator
20. Low oil pressure indicator
21. High beam indicator
22. Instrument light switch
23. Map light in rear view mirror
24. Lighting master switch
25. Key type ignition switch, also energizing starting and warning lights circuits
26. Windshield wiper switch
27. Generator regulator
28. Generator
29. Starter motor
30. Ignition coil
31. Spark plugs
32. Ignition distributor
33. Low oil pressure indicator sending unit
34. Rear tail, stop and directional signal lights
35. Number plate light

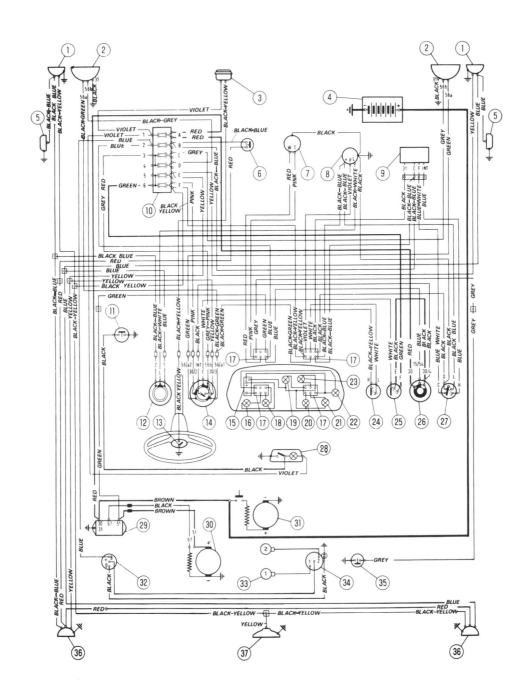

WD3 WIRING DIAGRAM FOR 500L SEDAN. FOR COLOUR CODE SEE FIG.WD2

1. Front parking and direction lights
2. High and low beam headlamps
3. Horn
4. Battery
5. Side direction indicators
6. Stop light hydraulic switch
7. Fuel gauge sending unit
8. Direction indicator flasher
9. Windshield wiper motor
10. Fuses
11. Push-button switch on driver's side door for light 28
12. Change-over switch for direction indicators
13. Horn push button
14. Change-over switch for outside lights and signal flasher
15. Fuel gauge
16. Low fuel indicator (red light)
17. Electric cable connections
18. No-charge indicator (red light)
19. Direction indicator repeater (green light)
20. Parking light indicator (green light)
21. Headlamp high beam indicator (blue light)
22. Low oil pressure warning light (red)
23. Instrument cluster light
24. Instrument cluster light switch
25. Outside light switch
26. Ignition switch controlling other circuits
27. Windshield wiper switch
28. Interior light, built in rear view mirror
29. Generator regulator
30. Generator
31. Starting motor
32. Ignition coil
33. Spark plugs
34. Ignition distributor
35. Low oil pressure indicator sending unit
36. Rear parking, stop and direction lights
37. Number plate light

Chapter 9 Rear suspension, wheels and tyres

Contents

General description ... 1	Rear suspension geometry ... 7
Removing the hub shaft ... 2	Shock absorber - checking ... 8
Hub bearings ... 3	Shock absorber - removal ... 9
Removing the rear suspension ... 4	Wheels ... 10
Suspension - overhaul ... 5	Tyres ... 11
Refitting the suspension ... 6	

Specifications

Wheels ... 12 x 3½ inch, pressed steel, disc

Tyres ... 125 – 12

Rear suspension dimensions
- Half-track (symetrical) Sedan ... 567.5 mm ± 1.5 mm
- Station wagon ... 565.5 mm ± 1.5 mm
- Rear suspension toe in ... $0° - 0° 25'$ (equivalent to 0 – 1/8 inch (0 – 4 mm)
- (checked with wheel vertical)
- Rear hub bearing rotates on torque under pre-load ... 0 - 36 lbf.ft (5.0 kg.cm)
- Spring free height: Sedan ... 221 mm (219)
- Station wagon ... 222 mm (220)
- (From car 141707 in brackets)
- Spring rate: Sedan ... 17 mm/100 kgf
- Station wagon ... 12 mm/100 kgf

Shock absorbers
- Interior diameter ... 27 mm
- Extended length: Rear ... 271 mm
- Front ... 335 mm
- Telescoped length: Rear ... 180 mm
- Front ... 212 mm
- Fluid filling: Rear ... 100 cm^3
- Front ... 130 cm^3
- Fluid. FIAT S.A.I. oil

Tightening torques
- Rear suspension mounting bracket to floor ... 29 - 36 lbf. ft (4 - 5 kg.m)
- Rear suspension pivot pin nuts ... 43½ - 50½ lbf.ft (6 - 7 kg.m)
- Flexible coupling to hub shaft as required to pre-load bearing
- Drive shaft to flexible coupling ... 20½ lbf. ft (2.8 kg.m)
- Wheel bolts: Sedan ... 32½ - 40 lbf.ft (4.5 - 5.5 kg.m)
- Station wagon ... 43½ - 50½ lbf.ft (6.0 - 7.0 kg.m)

Tyre pressures

	Sedan Crossply	Sedan Radial	Station wagon
Front ...	18 (1.3)	16 (1.1)	17 (1.2)
Rear ...	23 (1.6)	23 (1.6)	27 (1.9)
Rear (full load) ...	27 (1.9)	27 (1.9)	30 (2.2)

lbf/in^2 (Kg/cm^2)

Chapter 9/Rear suspension, wheels and tyres

1 General description

1 The rear suspension is independent, of semi-trailing arm layout.
2 The station wagon is of the same pattern, but there are detailed differences due to the extra weight and larger brakes. Therefore the actual springs, the suspension arms, and the shock absorbers, are different.
3 The suspension can be dismantled quite readily. It is adjustable, and care must be taken that the wheel geometry is not upset, or the handling will be strange and rear tyre wear bad.
4 The shock absorbers, or dampers, are of the telescopic type. Those at the front are basically similar, but have different settings, and are longer.
5 In time the springs do settle. The rubber bushes and pads perish and get cut by abrasion. After extended mileages the complete suspension requires overhaul. The shock absorbers require more frequent attention.

2 Removing the hub shaft

1 Jack up the car, remove the wheel, and apply the handbrake. Make sure the car is well blocked up elsewhere in addition to the jack so it cannot fall. The nut on the hub is quite stiff: The handbrake will prevent the hub from turning whilst undoing it, but there will be quite a load on the car and it must be certain that the car cannot be dislodged from the jacks whilst working on the nut underneath.
2 Disconnect the drive shafts by removing the three bolts that hold their coupling sleeves to the flexible coupling at the rear of the hubs. Push the drive shafts to one side, taking out the little springs between them and the hub. Tie the shafts out of the way with some string.
3 Clean the area around the hub.
4 Take the split pin out of the nut on the hub shaft now exposed by the removal of the drive shaft. Undo the castellated nut. (17 mm spanner).
5 Take out the four bolts holding the brake drum to the hub carrier.
6 Release the handbrake.
7 Pull off the brake drum.
8 Clean out all dirt and brake dust from the hub area. Pull the flange on the hub shaft towards you. From behind the bearing catch the large flexible coupling which has a rubber bush inside a metal body, and the spacer ring, that come off the shaft.
9 If the shaft does not come, put an old nut on the shaft and hammer with a soft faced hammer.
10 Check that the rubber in the flexible coupling is not being pushed out due to age. See that the splines are in good condition and coat them with grease before reassembly.
11 Reassembly is done in reverse order. The spacer goes on before the flexible coupling, the splines greased. Special attention must be paid to setting the preload on the hub bearings. Temporarily fit two of the brake drum bolts to rig up a spanner to control the hub whilst tightening the nut on the shaft end.
12 The castellated nut at the far end of the hub shaft sets the bearing preload. There is also resilience in the bearing spacer.
13 Tighten the nut gradually, rotating the hub as this is done to allow the bearings to move into position.
14 Line up one of the two split pin holes in the shaft with the nut slits.
15 Now check the bearing preload. Put a bar across the hub flange, tied to the holes so that it sticks out equally either side. Hang a weight of 1 lb 4.3 inches out from centre. This should have the bearing just on the move. If it does not the bearing is too stiff. If it is too stiff fit a new resilient bearing spacer, as this must have crushed too much under the tightening load of the nut on the end of the shaft.
16 Refit the drive shaft, greasing its splines and the little spring.

3 Hub bearings

1 The hubs can be stripped after the shaft has been removed. If the hub bearings need repair it is best to unbolt their housing from the rear suspension arm. Undo the four nuts that secure it and the brake back plate to the end of the arm. Lift off the hub, and put back a nut to hold the brakes. If the state of the bearings is unknown partial stripping is possible by only removing the grease seal at the outer end, which will allow the state of the bearings to be seen.
2 Assuming a full overhaul is necessary, take out the seals, circlips, inner races, the roller bearings, and the resilient spacer.
3 Clean the bearings very carefully. Check the races are bright and free from any pitting, and the rollers shiny.
4 If the bearings need replacement, fit with them a new resilient spacer. Only take the outer races out of the hub if they are to be replaced. To do so they want to be drifted out straight, ideally with a large tube.
5 Smear the bearings with grease. Do not pack the hub very full; just coat it all over inside, and the spacer. Fit the seals lips inwards. Drive them in straight, and gently. The old ones should not be refitted, use new ones.
6 The bearing preload is reset when the shaft is replaced as described in the previous section.
7 If the bearings merely need repacking with grease as part of the 18,000 mile task it is best to do so by removing the outer seal. This enables some parts to be removed, and gives room to get at the others. Take out the race so freed, and the spacer. Wipe away the old grease, and smear in new. Fit a new seal, and reassemble.

4 Removing the rear suspension

1 Jack up the rear of the car, and support it by spreading the load onto firm blocks, and freeing the jack.
2 Now put the jack under the end of the suspension arm near the hub, and take some of the weight, so that the shock absorber is not fully extended.
3 The brakes must now be disconnected; Plug the hydraulic reservoir, and disconnect the flexible pipe from the metal pipe under the body as described in Chapter 7.5.
4 Unhook the handbrake pull off spring between the suspension arm and the lever on the brake back plate.
5 Take out the pin fixing the cable to that lever.
6 Unclamp the outer cable at the adjuster from the bracket on the suspension arm.
7 Undo the three bolts holding the drive shaft sleeve to the flexible coupling on the back of the hub. Pull aside the drive shaft and lift out the little spring.
8 Undo the nut on the bottom of the shock absorber, and take off the washer and rubber bush. Telescope up the shock absorber out of the way. There is the other bush that was above the suspension arm.
9 Lower the jack slowly. It may be necessary to use a block to do so in stages as the jack reached the end of its travel. The weight is now being released from the spring.
10 Lower away until the spring can be removed from its seat. Note the rubber seat at the top end to reduce noise.
11 Remove the nuts on the bolts through the suspension pivots in the brackets under the floor. Do not take the front bracket off the floor. The three bolts that hold it there are the adjusters for the rear suspension geometry, and the toe-in will be upset.
12 Drive out the bolts through the rubber bushes for the suspension arm pivots. Note where all the washers and shims come from, so that they go back in the same place, lest again the position of the rear suspension be upset.

5 Suspension - overhaul

1 Check the free height of the spring.
2 Check the condition of the spring rubber seat, the bump

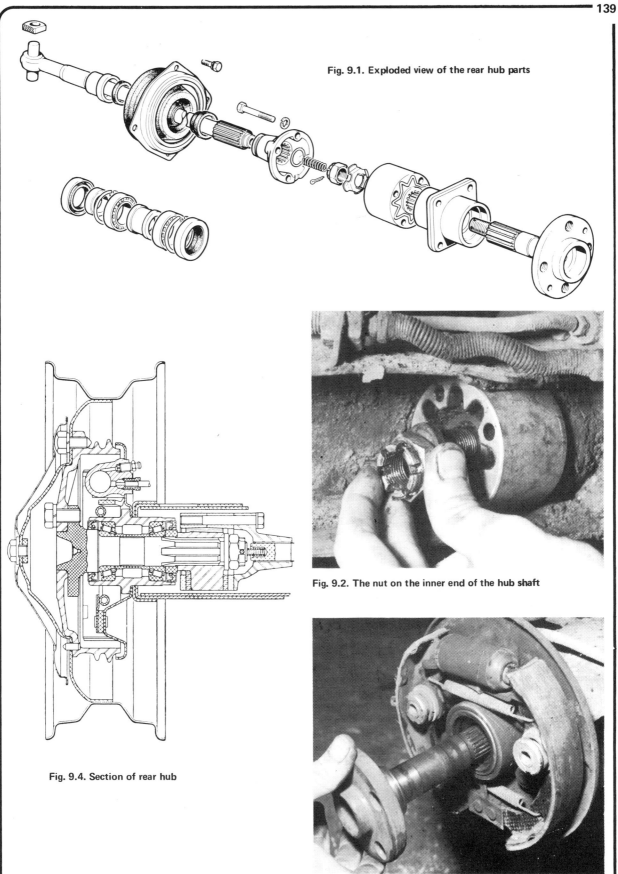

Fig. 9.1. Exploded view of the rear hub parts

Fig. 9.2. The nut on the inner end of the hub shaft

Fig. 9.4. Section of rear hub

Fig. 9.3. Removing the hub shaft

stop, and the rubber bushes.

3 The rubber bushes should almost certainly be replaced.

4 The condition of the bump stop is a good guide as to the spring sag. If the bump stop has been worked hard, as will be shown by its rubber, and the signs of rubbing where it seats on the suspension, the spring may be too short. About 15 mm is the maximum loss of free length that can be tolerated.

5 Check the suspension arm for any cracks or distortion.

6 Refitting the suspension

1 Put the suspension arm in position, with the new rubber bushes in their holes in the end.

2 Line up the holes with a screwdriver as a guide, then insert the bolts, with all the washers and shims.

3 Push up the suspension arm till in the normal laden position, and tighten the nuts on the pivots so that the rubber bushes are clamped in their working position.

4 Lower the suspension again, and fit the spring, with the rubber seat on top. This time take the weight with the jack under the arm. Jack up just enough for the shock absorber to be reconnected.

5 Refix the shock absorber, with the rubber bushes either side of the suspension arm, and their washers, and tighten the nut.

6 Reconnect and readjust the handbrake. (See Chapter 7.13).

7 Reconnect the hydraulic brake pipe and bleed the brakes. (See Chapter 7.4).

8 Couple up the drive shaft, greasing the splines and the little spring inside.

9 Check the rear suspension toe-in. See next section.

7 Rear suspension geometry

1 If the rear suspension is set up wrong the wheels will be crooked. This could give dangerous handling, the car reacting violently to the steering. It could also give rapid tyre wear.

2 The suspension is set by FIAT by putting the suspension arm in a jig to establish what shims are needed to set the width at which the wheels run: the track. The half-track is given in the specification: Each wheel must be the specified distance out from centre. There is some tolerance, but each wheel must err from the basic datum by the same error. If it has to be altered then it is done by shifting shims either side of the rubber bushes for the suspension arm pivots. It must be measured with the wheels vertical, as when heavily laden.

3 Unless the car has been damaged in an accident, the track adjustment should not be lost. But it is for the retention of the track that the shims must always go back in the same place when the suspension is dismantled.

4 The toe-in is more easily disturbed, and is more important than the track. It is also easier to adjust.

5 To measure the toe-in the car should be loaded till the wheels are vertical. This can be judged by viewing the suspension arm, and by hanging a plumb line down beside the wheel.

6 Rig up a toe-in measuring system as described in the next Chapter.

7 The toe-in is adjusted by slackening the three bolts securing the forward suspension pivot to the floor, and shifting it within the slop of the bracket on the mounting bolts.

8 The required toe-in of 10' corresponds to 5.5 mm at a distance of 1.84 metres from the wheel centre. 1.84 m is the car wheel base.

Fig. 9.5. To undo the rear suspension the pin through the bush must be removed, not the three bolts holding the bracket to the floor.

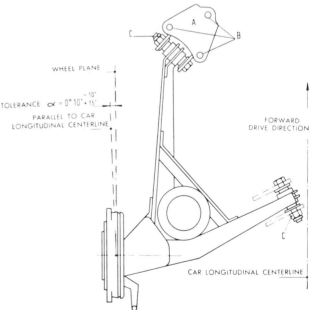

Fig. 9.6. The toe-in of the rear suspension is set by moving the mounting plate, A after slackening the bolts through the holes, B.

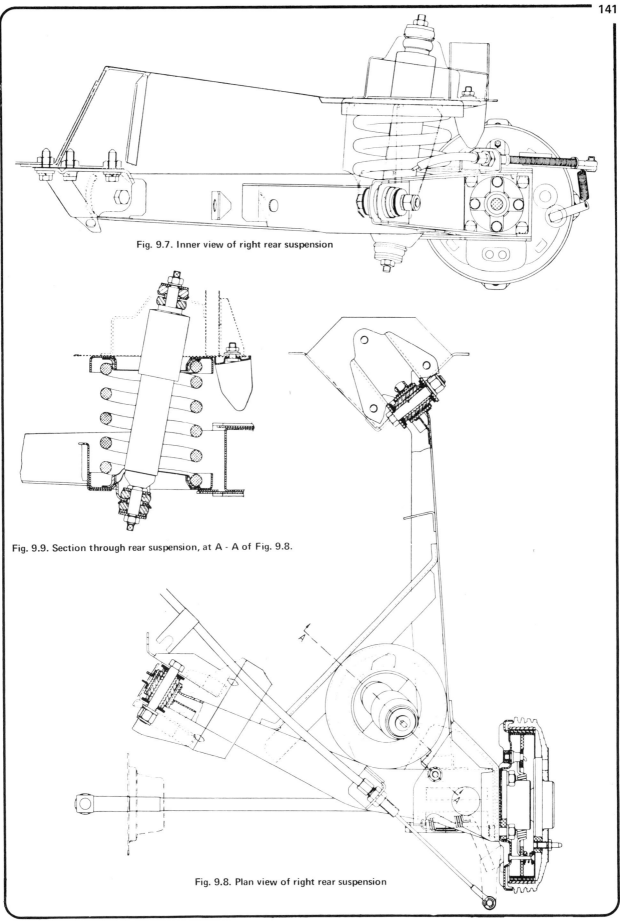

Fig. 9.7. Inner view of right rear suspension

Fig. 9.9. Section through rear suspension, at A - A of Fig. 9.8.

Fig. 9.8. Plan view of right rear suspension

8 Shock absorber - checking

1 Rather unusually, these can be dismantled. Spare parts are catalogued, though availability is unlikely to be good.
2 Shock absorbers last about 40,000 miles which is a long time compared with many cars. This is due to this car's light weight.
3 Their filure by becoming weaker, comes on gradually, and is difficult to detect. Leaking is a sure sign of failure.
4 Bounce vigorously one end of the car, timing the bouncing to be in phase with the car's spring bounce. When a good movement has been going, release the car. It should go once only past the normal position: The next time stopping at the static height.
5 Do not replace single shock absorbers unless the failure in an unusual one after a short mileage: Replace them as pairs, front or rear.
6 The front shock absorbers are longer than the rear, so are not interchangeable. Those on the station wagon have different valve settings.
7 Overhaul to renew worn shock absorbers is unlikely to be successful. However, as these ones can be dismantled, the fluid can be replaced, and one cause of failure rectified.

9 Shock absorber - removal

1 Remove the shock absorbers from the car by taking off the nuts, washers and rubber bushes at the bottom from the suspension, and at the top on the body.
2 Always keep the shock absorbers upright, so that the air is kept in the top. Before fitting new or refilled ones, work them a few times through their complete stroke to pump air to the outside.
3 The movement should be smooth, quite easy if moved slowly, but stiff to faster movement.
4 To strip the shock absorber, extend it fully. Clean it all very thoroughly. Undo the ring at the top of the cylinder, then pull out the rod and inner cylinder. The valve holder at the bottom of the inner cylinder is a push fit. Pump out all the old fluid. Refill with a measured amount of shock absorber fluid (see the specification). If you put in too much the shock absorber cannot compress fully.

10 Wheels

1 Wheels can get bent when kerbs are "nudged". During maintenance tasks, spin them round and check they are not out of true by more than about 1/10 inch (2.5 mm).
2 Include in your car cleaning programme washing the backs of the wheels. Always do this before having them balanced.
3 The wheels require repointing occasionally, either after fitting new tyres, or after some two years use. Rusting is liable to start where the wheel rim is joined to the disc. If this becomes severe the wheel may fail.

11 Tyres

1 The tyres are especially important to safety on a rear engined car. The recommended rear pressures are high to improve stability.
2 If radial ply tyres are fitted, and they are highly recommended, they should be fitted to all wheels.
3 Fitting tubeless tyres is best left to a tyre factory because they have the equipment to force the tyre into position before inflating it.
4 Changing tyres round to even out wear is not recommended. Apart from the large expenditure of buying five tyres in one batch, it masks any aberrations of wear which might be happening on one wheel.
5 If the tyres nearest the kerb side, that is the left tyres in Great Britain, wear faster than the one on the opposite side, this indicates the wheels are toe-ing in too much, and vice versa. This applies on the 500 to both front and rear.
6 If the treads wear in the centre before the edges this indicates over-inflation and vice versa.
7 It will pay to have the wheels balanced. Wheel out of balance gives shaking, which is unpleasant and causes much wear to the suspension and tyres.
8 Examine the tyres for cracks. Any cracks or bulges should be dealt with by a repairer: though seldom can they be cured. They are dangerous and illegal.

Chapter 10 Front suspension and steering

Contents

General description ... 1	Camber and caster adjustment ... 11
Front hubs - stripping ... 2	Front wheel toe-in ... 12
Front hubs - overhaul ... 3	Steering box - description ... 13
Front hub reassembly and adjustment .. 4	Steering box - adjustment ... 14
Removing track-rod ball joints ... 5	Steering box - removal ... 15
Stripping the front suspension ... 6	Steering box - stripping ... 16
Overhauling the king pins ... 7	Steering box - reassembly ... 17
Front spring - removal ... 8	Idler arm ... 18
Front spring - overhaul ... 9	Removing the upper steering column ... 19
Wishbone pivots ... 10	Fault finding ... 20

Specifications

Front spring leaves ... Main and four (five on station wagon) polyethylene interleave strips

Front suspension
- Front spring camber ... See text
- Front hub end float ... .001 - .004 inch (.025 - .100 mm)
- King pin bush reamed diameter ... 0.591 - .592 inch (15.016 - 15.043 mm)
- King pin new clearance ... .0006 - .0021 inch (.016 - .054 mm)
- King pin maximum permissible clearance .. .0079 inch (0.20 mm)
- King pin shims ... 19 sizes at 0.02 mm up to 2.80 mm)

Dimensions
- Wheel base ... 72.4 inch (1.84 m)
- Front track ... 44.1 inch (1.121 m)
- Toe-in ... 0 - 1/16 inch (0 - 2.0 mm)
- Wheel camber .. 1° ± 20' = 5 - 6 mm at rim
- Wheel caster ... 9° ± 1°
- Camber/caster adjustment shims ... .02 inch (0.5 mm)

Steering box (worm and sector)
- Clearance: drop arm shaft to eccentric bush: new ... 0.0 - .0017 inch (.042 mm)
- condemnation ... .004 inch (0.10 mm)
- Turning circle ... 28 ft 2 in (8.6 m)
- Steering wheel turns lock to lock ... 3

Tightening torques
- Wheel bolts (station wagon)... 43½ - 50½ lbf. ft (6.0 - 7.0 kg.m)
- (sedan) ... 32½ - 40 lbf. ft (4.5 - 5.5 kg.m)
- Suspension tap pivot bolt of king pin housing ... 40 - 43 lbf. ft (5.5 - 6.0 kg.m)
- bottom bolt of king pin housing... 29 lbf. ft (4.0 kg.m)
- Wishbone inner pivots: Inner pivots to body ... 29 lbf. ft (4.0 kg.m)
- Brake backplate to stub axle ... 14.5 lbf. ft (2.0 kg.m)
- Idler pivot pin nut ... 40 - 43 lbf. ft (5.5 - 6.0 kg.m)
- Drop arm to shaft ... 72½ - 79½ lbf. ft (10 - 11 kg.m)
- Steering box to body (and idler) ... 14½ - 18 lbf. ft (2 - 2.5 kg.m)
- Steering wheel to column ... 29 - 36 lbf. ft (4 - 5 kg.m)
- Ball joints to steering arms ... 18½ - 21½ lbf. ft (2.5 - 3.0 kg.m)

Chapter 10/Front suspension and steering

1 General description

1 The front suspension is independant, using one semi-elliptic leaf spring. Across the car the spring is mounted at two points. This mounting makes the spring twist into an 'S' shape when the car rolls on corners, so resisting this roll, and making a separate anti-roll bar unnecessary.
2 The steering is by a sector and worm box, with an idler arm and three piece track rod.
3 The front shock absorbers are longer, but otherwise similar to the rear, and are described in Chapter 9, as are the wheels and tyres.

2 Front hubs - stripping

1 Jack up. If part of more general overhaul of the front suspension, support the car fully, spreading the load and relieving the suspension.
2 Remove the road wheel.
3 Clean the hub near the bearing cap.
4 Prise off the bearing cap.
5 Undo the bearing nut. On the right side, after car number 043624 it is a left hand thread, (photo).
6 Take off the keyed washer.
7 Pull off the drum/hub assembly, (photo).
8 If it will not come easily, use a puller, though a very big one will be needed to reach out to the brake drum edge. It is possible on those models with a strap for the hub cap over the centre of the wheel to use the wheel as the puller. Refit the wheel, being careful not to scatter dirt into the bearing. Put a long bolt through the fitting for the hub cap, so that it pushes on the stub axle, and pushes off the hub.
9 Take out the small bearing race and rollers. Keep them clean.
10 Clean the stub axle and inner race of the larger bearing which will have been left behind on it. Check it for cracks or scores, particularly at its inner end.
11 If only stripped for repacking with grease, wipe away the old grease from both bearings and from inside the hub. Smear in the new grease. Whilst being liberal, do not fill the hub full.

3 Front hub - overhaul

1 Remove the oil seal.
2 Take out both sets of bearing rollers.
3 Wash the bearings thoroughly.
4 Check the rollers are bright and shiny.
5 Check that the inner and outer races are without blemish.
6 If any of the bearings are not faultless, they must be replaced. Press out the outer races, after removing the circlip outside the larger one. Be careful to keep them straight. A large pipe can make a useful drift.
7 Take the inner race of the larger bearing off the stub axle.

4 Front hub reassembly and adjustment

1 If new bearings are being fitted, press the new outer races into the hub, and the large inner race onto the stub axle.
2 Line the hub with grease. Grease the two roller bearings ready for assembly. Work the grease well into the rollers.
3 Put the large roller bearing into the hub, and fit its circlip.
4 Carefully put in a new grease seal, lips inwards.
5 Smear the stub axle with grease.
6 Lift the hub/drum assembly into the stub axle, carefully centering it so that the grease seal does not strike the axle.
7 Hold the hub with one hand to keep it straight: Put the small roller bearing onto the stub axle, and push it into the hub.
8 Fit the keyed washer to the stub axle, followed by the nut (preferably a new one if the self locking type).
9 Tighten the bearing nut on the stub axle. Whilst tightening it, rotate the hub to and fro so that the bearings can roll into their proper position.
10 The self locking type of bearing nut should be tightened only to a torque of 5.1 lb ft (0.7 kg m). Then it should be slackened 30°. This angle is ½ one flat of the nut. Make a mark on the washer behind the nut with a punch, level with a corner of the nut. Then slacken till the next flat of the nut is by the mark. Now lock the nut by staking it with a chisel into the groove in the stub axle.
11 The castellated nut should be tightened to a torque of 22 lb ft (3.0 kg m), and then slackened 60o. This is a full flat. So for this type make a mark on the washer behind the nut with a punch, just by a corner, and slacken the nut till the next corner is level with the mark; or such less angle that will allow the split pin to be fitted. Fit the split pin, bending one tab down towards the washer, and the other over the end of the stub axle.
12 When set like this the hub should be left with the required end float of .001 - .004 in (.025 - .100 mm).
13 It is unusual to have left hand threads on the right of a car. Normally it is the other way round, so that the rotation will tend to tighten the nut. This arrangement is on this car so that should the bearing seize, the wheel will not lock. Instead the inner race can be dragged round on the stub axle.
14 Put a little grease into the bearing cap, and tap it gently into place, keeping it straight as it goes home.

5 Removing track-rod ball joints

1 The ball joints on the two outer track-rods are in pairs. One has a left hand thread where it screws into the track-rod, and its mate a right hand one. So the track-rod is a turnbuckle for adjusting the toe-in. The ball joints on the centre tie-rod are integral with it.
2 If any play at all develops in the ball joints, they must be replaced. If the rubber boot fails, the ball will soon after.
3 To remove a ball joint from its seat in the steering, drop, or idler arms is sometimes difficult, as they are in tapered seats, and the pin of the ball joint tends to seize in place.
4 Remove the nut from the end of the pin.
5 If possible get a 'ball joint separator', and use this special tool to press the joint out of the arm.
6 If no tool is available the pin must be bounced out of its seat by impact. It wants to be hit from both sides simultaneously. Put one hammer against the arm beside the pin, then hit the opposite side a hard blow with another hammer.
7 If this is not successful wet the pin with easing oil such as WD 40, and allow it time to soak in. Wiggle the steering wheel to and fro to loosen the pin. Then try again. Usually putting the nut on the end of the thread and hitting that does no more than wreck the thread. Also take care all the hammering does not damage anything else.
8 When fitting ball joints to the arm smear a little grease to the taper of the pin. Ensure the rubber dirt-excluding boot is in good condition.
9 After fitting, or even refitting an old, ball joint, check the toe-in as described in section 12.

6 Stripping the front suspension

1 The stub axle and king pin housing assembly can be removed quite simply.
2 Jack up the front of the car. Transfer the weight to blocks under the body, spreading the weight broadly, and making sure that the car is very secure.
3 Move the jack to the outer end of the spring, and take some of the weight on this, so that the shock absorber is telescoped about an inch or a bit more.
4 Remove the wheel, and dismantle the hub bearing as described in section 2.
5 Take off the brake back plate. This is held by one nut on a stud through the steering arm, and another to the rear of the king pin. Lift off the back plate, still with the brake shoes on it,

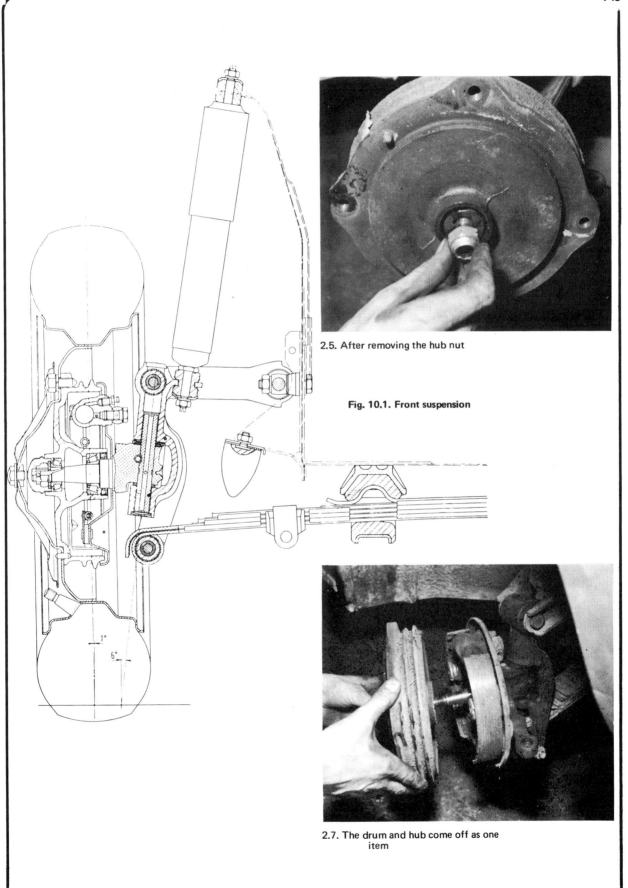

2.5. After removing the hub nut

Fig. 10.1. Front suspension

2.7. The drum and hub come off as one item

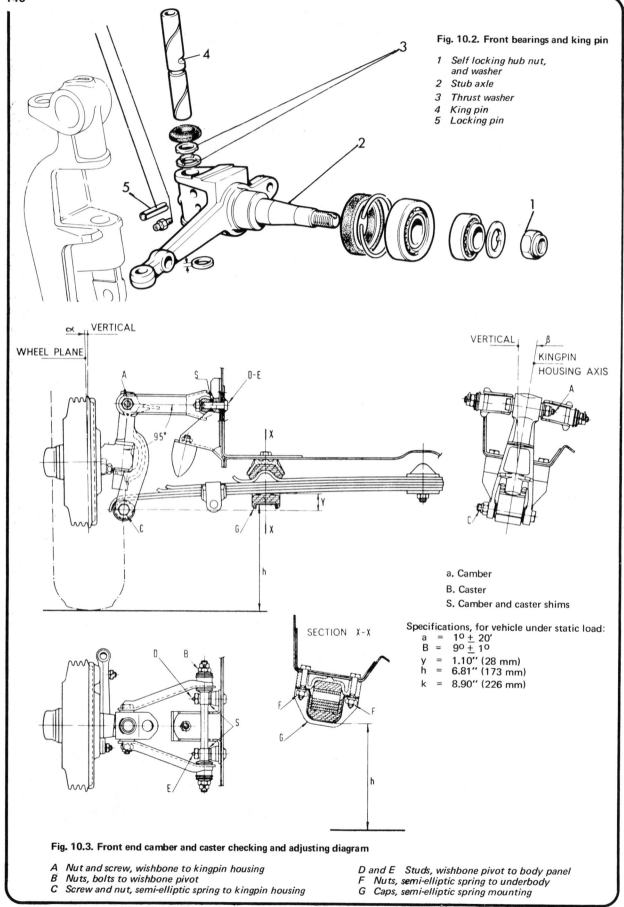

Fig. 10.2. Front bearings and king pin

1 Self locking hub nut, and washer
2 Stub axle
3 Thrust washer
4 King pin
5 Locking pin

a. Camber
B. Caster
S. Camber and caster shims

Specifications, for vehicle under static load:
a = 1º ± 20'
B = 9º ± 1º
y = 1.10'' (28 mm)
h = 6.81'' (173 mm)
k = 8.90'' (226 mm)

Fig. 10.3. Front end camber and caster checking and adjusting diagram

A   Nut and screw, wishbone to kingpin housing
B   Nuts, bolts to wishbone pivot
C   Screw and nut, semi-elliptic spring to kingpin housing
D and E   Studs, wishbone pivot to body panel
F   Nuts, semi-elliptic spring to underbody
G   Caps, semi-elliptic spring mounting

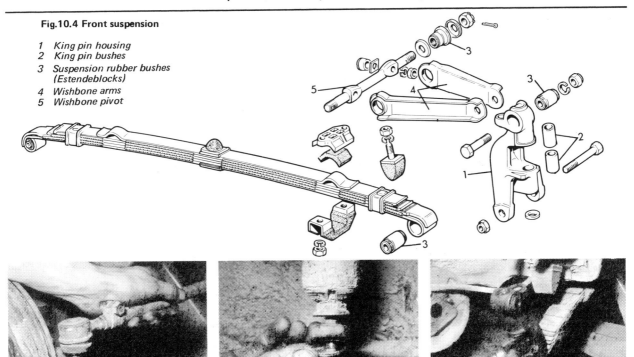

Fig.10.4 Front suspension

1 King pin housing
2 King pin bushes
3 Suspension rubber bushes (Estendeblocks)
4 Wishbone arms
5 Wishbone pivot

6.6. Undo the ball joint from steering arm. It may need a judicious blow with a hammer to jolt it out

6.7. Disconnect bottom end of shock-absorber, as it is acting as a rebound stop. Push it up out of way, but don't loose washers or bushes

6.8. There is no need to disconnect brake lines, but take care of pipe, when disconnecting suspension pivot bolts through king pin housing bushes

and the hydraulic pipe connected. Take care the hydraulic pipe is not strained by pulling or twisting. Tie the back plate up out of the way, or prop it on a pair of sticks, leaning against the mudguard.

6 Undo and remove the ball joint from the steering arm as described in the previous section, (photo).
7 Disconnect the bottom end of the shock absorber from the wishbone, the top suspension control arm. Retrieve the lower rubber bush. Telescope the shock absorber as far as it will go upwards. Remove the other washers and rubber bush from the thread at its bottom, (photo).
8 Remove the nut from the pivot bolt through the spring eye at the bottom of the king pin housing. Drive out the bolt. (photo)
9 Lower the jack under the spring. Pull the king pin housing clear of the spring, so that it is hanging from the wishbone at the top.
10 Repeat for the bolt through the top of the king pin housing and the wishbone. Now pull the whole king pin housing and stub axle assembly away.
11 Note all the washers, and where they came from.
12 Reassembly is done in the reverse order. If new rubber bushes (known to FIAT as 'Estendeblocks') are being fitted press these in first.
13 When the king pin housing has been fitted to the wishbone, raise it to the laden position (see Fig.10.2). Tighten the bolt through the pivot at this angle, so that the rubber bush is normally in its relaxed position, and can flex either way as the suspension moved up and down. If tightened hanging free they will be twisted too far when the suspension goes up on a bump.
14 In all the front suspension pivots are rubber bushes. The phrase is used: 'Press in a new bush'. Presses must be made by ingenuity. If an electric drill is available, its stand makes a reasonable one; quick but not very strong. A stronger one is to use a long bolt, with large washers made up to the appropriate size to draw the bush along.
15 Whenever the suspension has been stripped, check the toe-in after reassembly. See section 12.

7 Overhauling the king pins

1 Remove the king pin and stub axle assembly as described in the previous section. Only remove the king pin from its housing if the bushes are in need of renewal. (See specifications).
2 The king pin is pegged to the stub axle. To remove it hammer out the pin with a punch. Prise out the disc peened into the bottom of the king pin housing closing the bottom bush.
3 Press out the rubber bush from the top of the king pin housing. Drive out the king pin from above through the space vacated by the rubber bush. If the disc at the bottom could not be removed, provided it has been loosened, it can be driven out with the king pin.
4 As the king pin comes out of its housing and the stub axle is removed, note the position of all the washers. There is a load carrying washer and rubber dirt seal above. Below is a spacer.
5 The king pin replacement kit should include a pin for locking the stub axle to the king pin, and all the washers and seals needed but there are many different thicknesses of spacers.
6 Take the old bushes out of the housing. If difficulty is found in pressing them out they can be collapsed by cutting with a hack saw downwards to split them.
7 Press in the new bushes.
8 Now ream the bushes to the size in the specification. This must be done very carefully to get the exact size, and it is done straight through the bushes as a pair so that they are reamed in line.
9 For reassembly grease the king pin and bushes. Get the stub axle into place in the housing. Put all the washers in place, holding them with grease.
10 Fit the new king pin, lining up its groove with the hole in the stub axle for the locking pin.
11 Hammer in the locking pin firmly. Fit the disc to the bottom bush, and peen it into place, being careful not to damage the disc.

8 Removal of a front spring

1 If the spring is being removed as part of a general overhaul of the suspension, remove the king pin housing as described in section 6.
2 But if only the spring is being removed, much less work is necessary. Jack up and lower the car onto firm supports under the body, with the load well spread.
3 Put the jack under one end of the spring, and take the load of the spring off the king pin housing.
4 Remove the nut from the pivot bolt through the spring eye and bottom of the kingpin housing. Drive out the bolt and retrieve all washers, noting where they came from.
5 Lower the jack.
6 Repeat for the other side.
7 Remove the two nuts from the two clamps holding the spring to the car floor, supporting the spring so that it does not fall, (photo).

9 Overhaul of the front spring

1 Before removing the spring as described in the previous section, check the camber of the spring leaves when laden according to the dimensions given in Fig. 10.3.
2 Having removed the spring, undo the clips and centre bolt holding the leaves together.
3 Clean and lay out all the leaves and interleaf plastic sheets.
4 Examine the rubber bushes in the spring eyes, and if soggy, perished or torn press the old ones out. (See section 6.14)
5 Individual leaves can be replaced if cracked or badly worn, but if the third leaf is bad, this one is not available as a spare, and the whole spring must be replaced. New plastic inter leaf sheets, and rubbers for the mounting and clips will probably be needed. The leaves should be clean, and free of paint.
6 If the leaves have settled, lowering the car, but are otherwise in good order, their camber can be reset. A local engineering firm should be able to do this.

10 Wishbone pivots

1 The 'wishbone', the top suspension control arm, is made in two halves; thus its name. It is mounted on rubber bushes ('Estendeblocks') on a pin bolted to the body side.
2 The bushes must be replaced if the rubber is perished, torn, or worn. The wishbone must be held by them centrally on the pin, and with only the designed amount of resilience in the bush. If too soggy, steering will be spoiled.
3 It is assumed this work is part of a more general suspension overhaul, and that the king pin housing has already been removed as described in section 6.
4 Remove the nuts on the two ends of the mounting pin.
5 Take off the washer, and pull off each half of the wishbone.
6 If the mounting pin is in good order, leave it in place but if removing it, note the shims on the two mounting points between the pin and the body. These set the camber and caster.
7 When reassembling, the nuts clamping the rubber bushes must be fitted with the wishbone up in the loaded position. See Fig.10.3

11 Adjusting camber and caster

1 Camber is the angle at which the wheel leans out sideways, the top being farthest out than the bottom (positive camber). Caster is the angle of the king pin so that the wheel has a self straigtening effect. The geometry is given in Fig. 10.3.
2 It is difficult to set these without proper gauges, but it should never need doing unless the car has been damaged.
3 The car must be laden so that the spring bracket at the front is 173 mm above the ground (dimension Y in Fig. 10.3. .
4 At the rear, the sump lowest point should be 156 mm from the ground, for the sedan. For the station wagon the rear body sill should be 226 mm above the ground.
5 Adjust camber by adding or taking out shims from both mounting points for the inner pivot pin for the wishbone.
6 To increase caster move shims from the rear stud to the front stud mounting the pivot pin, and vice versa. See item S in Fig. 10.3.

12 Front wheel toe-in

1 Garages have accurate gauges, so measure this readily, but even so there is possibility of error. Measuring at home is far less accurate, but time is at less of a premium, so it can be done often enough to get several consistent readings to know that the setting is correct.
2 If the car is old, and you have not set its steering yourself, the steering wheel and track rod may be off centre. It must be established that the steering box is in the centre of its travel when the wheels are straight, and that then the steering wheel is straight too.
3 Turn the steering wheel from lock to lock, and find its centre. Check the steering wheel is now straight. If not, unclamp the bottom of the steering column, removing the pinch bolt, slide the column off the steering box input shaft, and refit it straight. Now drive the car on the road, and see where the steering wheel is when travelling straight. This will tell you which track rod needs shortening, and which lengthening. Do this, altering each the same amount. This must be done accurately. Mark the track rods with chalk so the amount turned can be seen. Try only one turn at first. A little makes a lot of difference. The relative length of the two track rods can be measured too if the steering is a long way out.
4 Now check the toe-in as follows. This is all that will normally be needed, once the foregoing centralisation has been done.
5 Check the tyre pressures. Then set up on level ground the measuring bars as shown.
6 The measuring bars must be very straight. Useful ones are aluminium alloy strip, 1 inch by ¼ inch by 5 feet. On them put marks for the hub position 1 foot from one end. At the other end mark off a length twice the diameter of the wheel rim, 2 feet.
7 With the steering in the straight ahead position put the bars on blocks, or oil cans, or some such, so that they are at hub height. Arrange them to lie along the wheels, the marks for the hubs level with the hubs both vertically and horizontally. To make the bars firm, lean something against the bar opposite the hub so that it is pressed against the tyre on both sides.
8 Measure the distance across the gap between the two bars at their far end. Compare this with the distance apart at the other mark, 2 foot nearer. The difference is twice the toe-in. Make sure that it is toe-in you have got, not toe-out.
9 Take another reading. It will probably vary from the first. If so take another to get an average. These subsequent readings must be after moving the car forward. Do not move the car back, as the wheels will push on the steering the wrong way. As the wheel will not be perfectly true, readings must be taken at various parts of the rims to get a fair sample.
10 If a correction is needed, adjust both track rods if the steering is centralised. If it should be slightly off centre, adjust that one which will get it nearer straight. Mark the track rod so the amount turned can be seen. A quarter of a turn makes quite a difference. Then recheck.
11 When making the adjustment, unclamp the locknuts at both ends of the track rod. Turn the track rod. It is a turnbuckle with left hand threads at one end. When locking up after making the adjustments, push both sockets of the ball joints at the two ends of the track rod as far as they will go in the same direction on their balls. Then tighten up and by doing this the joint will have full articulation. Just check it can move freely afterwards to make sure.
12 Adjusting the tow-in is tedious, but if it is not correct, tyre wear is bad.

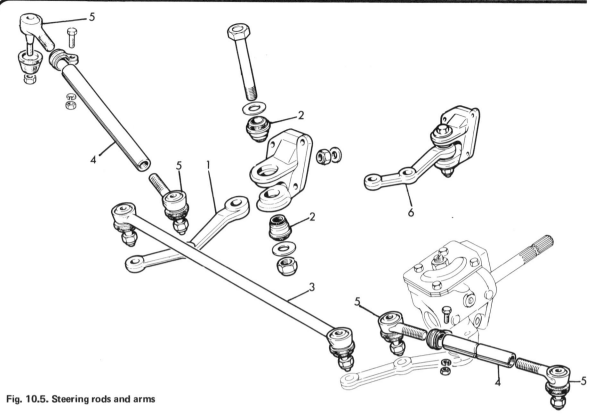

Fig. 10.5. Steering rods and arms

1 Idler arm	3 Tie-rod	5 Ball joints in left and right hand thread pairs
2 Idler arm rubber bushes	4 Track-rods	6 Idler arm assembly

8.7. After taking out the pivot bolts through the spring eyes and king pin housing it only remains to take off the two nuts on the two spring mounting brackets

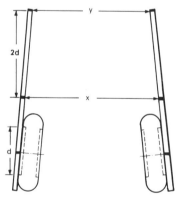

Fig. 10.6. Adjusting toe-in

$$\frac{x - y}{2} = \text{Toe-in}$$

13 Steering box - general description

1 The steering column is connected to a steering box, whose drop arm is connected direct by one track rod to the nearest wheel, and also works a central tie rod joined at its far end to an idler arm on the opposite side of the car to the steering box. From the idler arm another track rod goes to the far wheel.
2 The track rods go out at each side to ball joints on the steering arms on each stub axle.
3 The steering box is fully adjustable. Provided it is not damaged by such a thing as a blow or high stress as in an accident, it should easily last the life of the car. Wear in all components can be adjusted. The only likely requirement is to replace the eccentric bush for the drop arm shaft below the sector, and at the same time the oil seals.

14 Adjustments of the steering box

1 Adjustment of the steering box can be done in place. If a lot of other work is to be done, then it is worth removing it.
2 Our own photographs have been taken on the bench for clarity.
3 The most likely adjustment needed is the eccentric bush for the drop arm shaft. This eccentric bush can be moved by removing the setscrew holding the serrated plate to the bottom of the steering box housing. Having removed the setscrew the serrated plate can be shifted over, to turn the bush. This adjustment moves the sector on the top of the drop arm closer to the worm on the steering shaft. If insufficient adjustment is available the plate must be moved round on the bush.
4 To do this remove the drop arm from the drop arm shaft. It is held by a self locking nut on the end of the shaft, and must then be withdrawn using a puller. With this out of the way the serrated plate can be taken off and turned on the eccentric bush to re-engage on other splines further round. See item 30 in Fig. 10.6. (photo).
5 The setting required is for the sector to be as tight as it can get without binding or put the other way, bring it in till all free play just disappears. There should be immediate movement of the drop arm shaft as soon as the steering column is turned. This adjustment must be done with the wheels in the straight ahead position.
6 There are two types of plate for adjusting the eccentric bush. Early ones have one securing bolt, later ones have two.
7 The end float in the drop arm shaft is corrected by a peg screwed into the top of the steering box. This is held by a locknut. Slacken the locknut. Screw in the peg until it is in light contact with the sector underneath the cover. Hold it there and tighten the locknut. Note this adjustment must also be done with the wheels straight. The need to do it can be detected sometimes by a clonking heard or felt in the steering wheel on bad bumps taken slowly. It can be felt in bad cases when pulling the drop arm from beneath the car, (photo).
8 The end float in the worm, which is the extension of the steering column in the box, is done by the castellated cap in the bottom of the box, (photo).
9 Take out the split pin. Screw the cap, using long nosed pliers opened wide, till all end float goes. But do not overtighten. There should be no stiffness.
10 Insert the split pin into whichever of the two holes lines up best.
11 This end float should not need resetting in service, but will need adjustment after the box is stripped.

15 Steering box - removal

1 To remove the steering box for major overhaul, proceed as follows:
2 At the bottom of the steering column just above the floor, remove the bolt pinching the steering column to the shaft into the steering box.

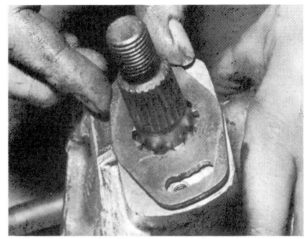

14.4. With the drop arm and dirt shield removed for the photo, the serrated adjuster for the eccentric bush on the drop arm shaft can be seen. The range of adjustment beyond the slot in the adjuster can be extended by stripping it like this and lifting it round a few splines

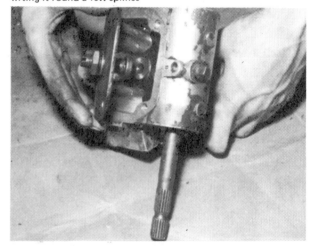

14.7. The cover off the steering box to show the adjusting screw for drop arm end float, and its locknut

14.8. The serrated adjuster for worm endfloat. There are two holes for the split pins, allowing fine adjustment

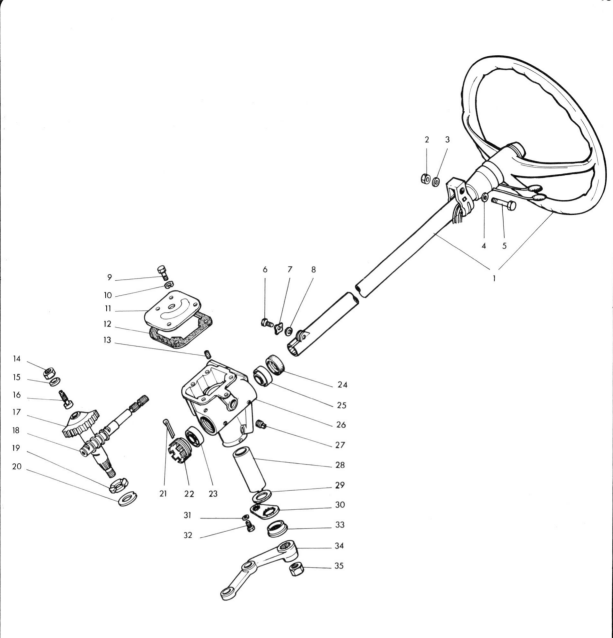

Fig. 10.7. Steering box and column

1. Steering wheel and column assembly
2. Nut
3. Toothed washer
4. Plain washer
5. Steering column bracket-to-instrument panel bolt
6. Steering column-to-worm locking screw
7. Lock plate
8. Plain washer
9. Cover screw
10. Toothed washer
11. Cover
12. Gasket
13. Pin
14. Nut
15. Plain washer
16. Sector adjusting screw
17. Worm sector and shaft assy.
18. Worm screw
19. Thrust washer
20. Shim
21. Split pin
22. Lower sleeve, bearing retainer and worm screw adjuster
23. Roller bearing
24. Worm screw bearing upper seal
25. Roller bearing
26. Steering gear housing
27. Oil filling and draining plug
28. Eccentric bush
29. Upper seal
30. Worm screw-to-sector lash adjusting plate
31. Toothed washer
32. Plate screw
33. Lower seal
34. Drop arm
35. Self-locking nut

3 Underneath the car remove the ball joints for the driver's side track rod, and the central tie rod, from the drop arm, as described in section 5.
4 Remove the three nuts holding the steering box to the floor of the car. Lift the steering box off.

16 Stripping the steering box

1 Stripping the steering box is straightforward.
2 Undo the bolt holding the drop arm to the drop arm shaft and pull it off.
3 Remove the screws holding the steering box top, and remove that.
4 Take out the sector attached to the drop arm. With it may come the eccentric bush. Between the sector and the casing is a thick thrust washer. Note the small peg that prevents the thrust washer from turning, sticking out from the body of the steering box. Also there are likely to be shims between the thrust washer and the steering box.
5 Remove the split pin from the castellated cap at the bottom of the worm shaft.
6 Unscrew the castellated cap. Long nosed pliers can be used for this if the special tool is not available. With the cap removed the worm should be tapped very gently on the end of the shaft where the steering column is attached. This will drive out the race for the bottom taper roller bearing. Once this is removed the worm with the two inner races can be removed. Note there is a seal inside the top race mounted in the steering box. This stops oil coming out upwards to the shaft. The seal for the bottom is the castellated cap.

17 Reassembly of the steering box

1 Reassembly of the box is the reverse process.
2 When inserting the worm check that the seal at the top is in good condition. To get in the bottom bearing's outer race, tap it in using something like a socket to drive it in straight.
3 Before fitting the thrust washer for the sector, put in shims as needed to get the centre line of the sector's teeth level with the centre line of the worm. Then put in the thrust washer, locating its cut out with the peg in the box.
4 An oil seal goes round the drop arm shaft at the bottom of the eccentric bush inside the serrated adjustment plate. After the plate is a metal shield with a dirt seal. On the early cars with the serrated plate having only one retaining bolt, the dirt seal is without a metal shield, the seal being held in by the drop arm.
5 Turn the steering box from one end of its travel to the other, and then find the middle. Then adjust the end float of the worm, the end float of the drop arm shaft, and then the eccentric bush for the meshing of sector and worm, all as described in section 14.
6 Fit the drop arm. It goes roughly straight as an extension of the direction of the steering column, but anyway it cannot be put on wrong as there is a master spline. The splines of drop arm and shaft are tapered.
7 Fill the box with SAE 90 EP gear oil to the level of the plug.

18 Idler arm

1 The idler arm is a relay lever on the opposite side of the car to the steering box. It transfers the steering effort from the centre tie rod to the outer track rod.
2 It is mounted on the car body by three studs in a similar manner to the steering box. The pivot is on rubber bushes. A certain amount of resiliance is designed into these, but if it becomes too sloppy the bushes must be replaced.
3 To remove the idler arm complete, first disconnect the steering knuckles for the centre and outer track rods. (See section 5). Note that the rubber bushes can be changed without removing the arm and its mounting from the car.
4 Now undo the three nuts holding the bracket for the pivot to the car body, and lift the assembly away.
5 To strip the pivot undo the large nut on the end of the pivot pin. Withdraw the pin, with washers either end, and the two bushes.
6 When reassembling the point should be noted that the rubber bushes must be tightened when the steering is in the straight ahead position. The torque for this is only approximately half that of the nut holding the drop arm onto the steering box: If you have not got a torque spanner only tighten this nut sufficiently to squeeze the rubber up firmly.

19 Removing the upper steering column

1 The column can be removed either with the steering wheel and all switches in place, or these can be removed first.
2 The steering wheel is removed by prising out the horn button, and removing the nut on the top of the column.
3 Disconnect the leads to the switches at their connectors, near the column. If the car is old and the colours have faded, they will need marking.
4 Undo and take off the bolts pinching the bottom of the upper steering column to the shaft into the steering box.
5 Remove the bolt holding the column to its bracket on the scuttle.
6 Twist the column round and pass it out of the door.

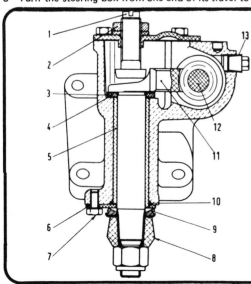

Fig. 10.8. Section of steering box

1 Drop arm shaft end float adjuster
2 Lock nut
3 Thrust washer
4 Shim
5 Eccentric bush
6 Bush adjusting plate
7 Its lock screw
8 Drop arm
9 Drop arm shaft lower seal early type
10 Upper seal
11 Sector
12 Worm
13 Oil filler and level plug

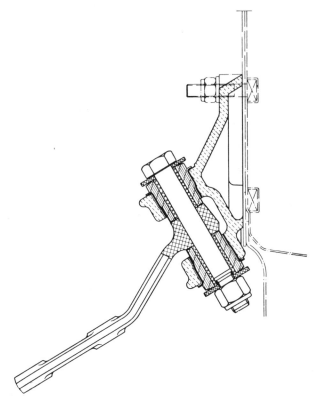

Fig. 10.9. Idler arm and mounting

Section 20 Fault finding is given overleaf

20 Fault finding

Symptom	Reason/s	Remedy
STEERING FEELS VAGUE, CAR WANDERS AND FLOATS AT SPEED		
General wear or damage	Tyre pressure uneven	Check pressures and adjust as necessary.
	Shock absorber worn	Test, and replace if worn.
	Steering gear ball joints badly worn	Fit new ball joints.
	Suspension geometry incorrect	Check and rectify.
	Steering mechanism free play excessive	Adjust or overhaul steering mechanism.
	Front suspension and rear suspension pick-up points out of alignment	Normally caused by poor repair work after a serious accident. Extensive rebuilding necessary.
	Mixed radial and crossply tyres	Fit all radial.
STIFF AND HEAVY STEERING		
Lack of maintenance or accident damage	Tyre pressure too low	Check pressures and inflate tyres.
	No oil in steering gear	Top up steering gear.
	No grease in steering and suspension ball joints	Clean nipples and grease thoroughly.
	Front wheel toe-in incorrect	Check and reset toe-in.
	Suspension geometry incorrect	Check and rectify.
	Steering gear incorrectly adjusted: too tight	Check and re-adjust steering gear.
	Steering column badly misaligned	Determine cause and rectify (usually due to bad repair after severe accident damage and difficult to correct).
WHEEL WOBBLE AND VIBRATION		
General wear or damage	Wheel nuts loose	Check and tighten as necessary.
	Wheels and tyres out of balance	Balance wheels and tyres and add weights as necessary.
	Steering ball joints badly worn	Replace steering ball joints.
	Hub bearings badly worn	Remove and fit new hub bearings.
	Steering gear free play excessive	Adjust and overhaul steering gear.
	Front spring weak or broken	Inspect and renew as necessary.
RATTLES	Suspension rubber bushes worn	Replace.
	Steering ball joints worn	Replace.
	Backlash in steering box	Adjust.

Chapter 11 Bodywork

Contents

General description 1	Back and side fixed window glasses 9
Maintenance - bodywork 2	Door windows 10
Maintenance - upholstery and floor coverings.. 3	Door ventilators 11
Minor body repairs 4	Door lining trim panels 12
Major body repairs 5	Door rattles - tracing and rectification 13
Body corrosion 6	Door hinges 14
Windscreen - glass removal 7	Door and luggage locker lid fit 15
Windscreen - glass replacement... 8	Emergency access to luggage locker 16

1 General description

1 The body is a simple construction, so without many corners and seams where corrosion may occur.
2 The body is fabricated by welded pressings. The power unit and suspension are bolted direct to it without any other framing.

2 Maintenance - bodywork

1 The FIAT 500 is a particularly easy car to keep clean due to its shape. The general condition of a car's bodywork is the one thing that significantly affects its value. Maintenance is easy but needs to be regular and particular. Neglect, particularly after minor damage, can lead quickly to further deterioration and costly repair bills. It is important also to keep watch on those parts of the car not immediately visible, for instance, the underside, inside all the wheel arches and the engine compartment.
2 The basic maintenance routine for the bodywork is washing - preferably with a lot of water, from a hose. This will remove all the solids which may have stuck to the car. It is important to flush these off in such a way as to prevent grit from scratching the finish. The wheel arches and underbody need washing in the same way to remove any accumulated mud which will retain moisture and tend to encourage rust. Paradoxically enough, the best time to clean the underbody and wheel arches is in wet weather when the mud is thoroughly wet and soft. In very wet weather the underbody is usually cleaned of large accumulations automatically and this is a good time for inspection.
3 Periodically it is a good idea to have the whole of the underside of the car steam cleaned, engine compartment included, so that a thorough inspection can be carried out to see what minor repairs and renovations are necessary. Steam cleaning is available at many garages and is necessary for removal of accumulations of oily grime which sometimes cakes thick in certain areas near the engine and transmission. The facilities are usually available at commercial vehicle garages but if not there are one or two excellent grease solvents available which can be brush applied. The dirt can then be hosed off.
4 After washing paintwork, wipe it with a chamois leather to give an unspotted clear finish. A coat of protective wax polish will give added protection against chemical pollutants in the air. If the paintwork sheen has dulled or oxidised, use a cleaner/polisher combination to restore the brilliance of the shine. This requires a little more effort, but is usually caused because regular washing has been neglected. Always check that door and ventilator opening drain holes and pipes are completely clear so that water can drain out.
5 Bright work should be treated the same way as paintwork. Windscreens and windows can be kept clear of the smeary film which often appears if a little ammonia is added to the water. Never use any form of wax or chromium polish on glass.

3 Maintenance - upholstery and floor coverings

1 Floor mats should be brushed or vacuum cleaned regularly to keep them free of grit. If they are badly stained remove them from the car for scrubbing or sponging and make quite sure they are dry before replacement. Seats and interior trim panels can be kept clean by a wipe over with a damp cloth. If they do become stained (which can be more apparent on light coloured upholstery) use a little liquid detergent and a soft nailbrush to scour the grime out of the grain of the material. Do not forget to keep the head lining clean in the same way as the upholstery. When using liquid cleaners inside the car do not over-wet the surfaces being cleaned. Excessive damp could get into the seams and padded interior causing stains, offensive odours or even rot. If the inside of the car gets wet accidentally, it is worthwhile taking some trouble to dry it out properly, particularly where carpets are involved. Do NOT leave oil or electric heaters inside the car for this purpose.

4 Minor body repairs

1 A car which does not suffer some minor damage to the bodywork from time to time is the exception rather than the rule. Even presuming the gatepost is never scraped or the door opened against a wall or high kerb, there is always the likelihood of gravel and grit being thrown up and chipping the surface, particularly at the lower edges of the doors and sills.
2 If the damage is merely a paint scrape which has not reached the metal base, delay is not critical, but where bare metal is exposed action must be taken immediately before rust sets in.
3 The average owner will normally keep the following 'first aid' materials available which can give a professional finish for minor jobs:

a) Matching paint in liquid form - often complete with brush attached to the lid inside. (Aerosols are extravagant. Spraying from aerosols is generally less perfect than the makers would have one expect).
b) Thinners for the paint (for brush application) (like Belco).
c) Cellulose stopper (a filling compound for small paint chips).
d) Cellulose primer (a thickish grey coloured base which can be applied as an undercoat in several coats and rubbed down to give a perfect paint base).
e) Proprietary resin filler paste (for larger areas of in-filling).
f) Rust-inhibiting primer (such as zinc plate).
g) "Wet or dry" paper grades 220 and 380.

4 Where the damage is superficial (i.e. not down to the bare metal and not dented) fill the scratch or chip with stopper sufficient to smooth the area, rub down with paper and apply the matching paint.

5 Where the bodywork is scratched down to the metal, but not dented, clean the metal surface thoroughly and apply the primer (it does not need to be a rust-inhibitor if the metal is clean and dry), and then build up the scratched part to the level of the surrounding paintwork with the stopper. When the primer/stopper is hard it can be rubbed down with "wet or dry" paper. Keep applying primer and rubbing it down until no surface blemish can be felt. Then apply the colour, thinned if necessary. Apply as many coats and rub down as necessary.

6 If more than one coat of colour is required rub down each coat before applying the next.

7 If the bodywork is dented, first beat out the dent as near as possible to conform with the original contour. Avoid using steel hammers - use hardwood mallets or similar and always support the back of the panel being beaten with a hardwood or metal 'dolly'. In areas where severe creasing and buckling has occurred it will be virtually impossible to reform the metal to the original shape. In such instances a decision should be made whether or not to cut out the damaged piece or attempt to re-contour over it with filler paste. In large areas where the metal panel is seriously damaged or rusted, the repair is to be considered major and it is often better to replace a panel or sill section with the appropriate part supplied as a spare. When using filler paste in largish quantities, make sure the directions are carefully followed. It is false economy to try and rush the job as the correct hardening time must be allowed before stages or before finishing. With thick application the filler usually has to be applied in layers - allowing time for each layer to harden.

8 Sometimes the original paint colour will have faded and it will be difficult to obtain an exact colour match. In such instances it is a good scheme to select a complete panel - such as a door, or boot lid, and paint the whole panel. Differences will be less apparent where there are obvious divisions between the original and repainted areas.

9 Do not expect to be able to prepare, fill, rub down and paint a section of damaged bodywork in one day and expect good results. It cannot be done. Give plenty of time for each successive application of filler and primer to harden before rubbing it down and applying the next coat.

10 Do not think that it is necessary, or even desirable to spray the paint on. To get a satisfactory result requires good equipment and experience. Yet a coachbuilder's finish is easily got by brush painting. The secret is the preparation and final polishing. All undercoats must be rubbed down with wet "wet or dry" paper of grade about 220, to remove all traces of the original damage, the edge of the filler, and the brush marks in the undercoat. Paint on a coat of top coat, brushing it out with criss-cross brushing to get it even. Brushing type thinners must be used as paint prepared for spraying will dry too fast, and go tacky before it is brushed smoothly out. This coat of the top paint can not be the final one as it cannot be put on thick enough without risk of weeping. Some areas will be so thin the undercoat will be visible immediately after painting. Anyway a good thickness of paint is needed for the polishing. So allow this coat to harden for 24 hours. Then lightly rub it down with wet "wet or dry" of about 350 grade. It must be rubbed mat all over for the final coat to key in, and to get rid of brush marks. Then put on the top coat. Take great care with this. Dry weather, without any wind to stir up dust is needed. Again, when hard after 24 hours, rub down very gently with grade 400 wet "wet or dry" just enough to take off the tops of the brush marks, and any blobs of dust. Then polish up with "rubbing compound", and abrasive paste. Then give the paint a few days to really harden, and polish with car polish; a cleaner or restorer, not a wax. Silvo metal polish is excellent for this. Incidentally Silvo can often be used for polishing out minor scratches that have marked the paint but not got down to the undercoat.

11 If you do apply the paint by spraying, then the guidance for rubbing down is still the same, but the job easier. The "orange peel" effect of one coat must be removed before the next coat is applied. Also polishing will be needed after the final coat, preceded by the gentle rub to get off the top of its lumps.

5 Major body repairs

1 Where serious damage has occurred or large areas need renewal due to neglect it means certainly that completely new sections or panels will need welding in and this is best left to professionals. If the damage is due to impact it will also be necessary to check the alignment of the body structure.

2 If a body is left misaligned it is first of all dangerous as the car will not handle properly - and secondly, uneven stresses will be imposed on the steering, engine and transmission, causing abnormal wear or complete failure. Tyre wear will also be excessive.

6 Body corrosion

1 The ultimate scrapping of a car is usually due to rust, rather than it becoming uneconomic to renew mechanical parts.

2 The rust grows from two origins: From the underneath unprotected after the paint was blasted off by road grit; from inside, where damp has collected inside hollow body sections, without a chance to drain, and no rust protection to the metal.

3 The corrosion is particularly prone to start at welded joints in the body, as there are stresses left in after the heat of welding has cooled. These seams are also traps for the damp, and difficult to rustproof.

4 Salt on the roads in winter promotes this horror. It is hygroscopic; it attracts damp, so the car stays damp even in a garage. It is also an electrolyte when wet, so promotes violent corrosion. If the car has been used on salty roads it must be desalted as soon as possible. A couple of days rain after a thaw clears the roads, and then driving the car in the wet does this naturally. The damage is worse if a car is not used for some time after getting salty.

5 The 6,000 miles (or 6 months) task includes the checking of the underneath for rust. It should be done just before the winter, so that the car is prepared for its ravages whilst dry and free of salt. Then it wants doing in the spring to remove the winters damage.

6 The bodywork should be explored and all hollow sections found. Into these a rust inhibitor should be injected; aerosols like Supertrol 001 or of Di-Nitrol 33B are good. Areas easier of access can be wetted by the inhibitor bought more cheaply as a liquid and applied by paint brush. An example are the insides of the doors, which can be reached by removing the panel linings. If hollow sections are sealed, it pays to drill a hole, spray in the aerosol, and then seal with the underneath paint smeared over the hole.

7 The underneath needs painting, and where abraded by grit flung up by the wheels, this must be one of the special thick resilient paints. "Adup" bronze super seal is recommended. It is compatible with the Supertrol 001 and Di-Nitrol 33B inhibitors. So if these are put on first the two between them make a good job in getting into corners. The "Adup" underneath paint can be used as ordinary paint on sheltered areas. On those showered by

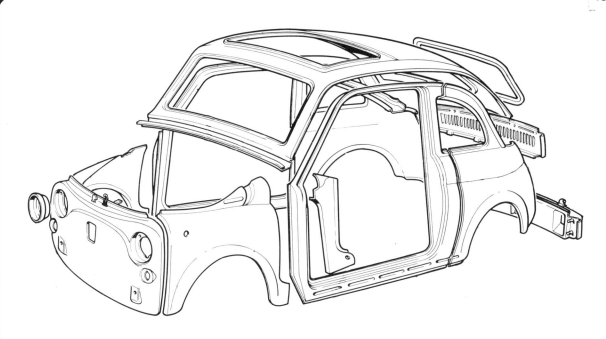

Fig. 11.1. The body is made up by welding these panels together. Sedan.

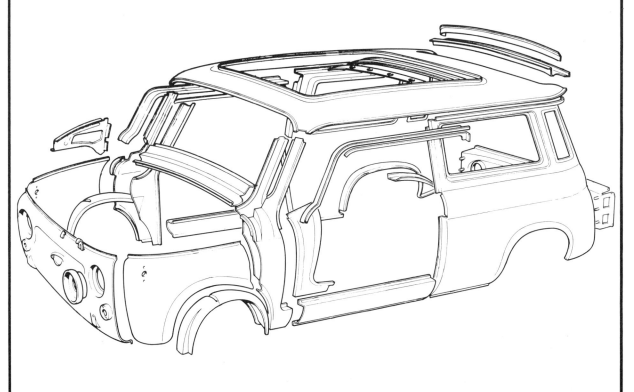

Fig.11.2 Station wagon welded panels

grit from the wheels, thick layers need to be applied. At the mudguards the underneath paint wants to be brought neatly round the edges so that these are protected, the corners being particularly vulnerable. Such protection can be extended to areas such as the hollow where the spare wheel sits.

7 Windscreen - glass removal

1 First check the maker's label on the glass. If the glass is laminated, and it will be if the car has been supplied for North America or Germany, amongst many markets, question whether you really must remove it. Toughened glass is, as its name implies, strong. Laminated glass is highly likely to crack.
2 To remove the glass, swing down the wiper blades clear of the windscreen.
3 Get an assistant (or two) outside, ready to catch the glass.
4 Get into the car; wearing thick gloves with gauntlets up the wrists for safety.
5 Push hard on the screen near one of the top corners. The screen will peel the rubber mount out of the body.

8 Windscreen - glass replacement

1 Assuming the old screen has shattered, cover up the demister ducts, and take out the broken remains of the old one. Clean the car with a vacuum cleaner.
2 Swing the wiper blades down forward clear of the screen opening.
3 Take off the old rubber surround.
4 Scrape off or wash away with petrol all old sealing compound round the aperture. If the paint is scratched, touch up all round the seating for the rubber seal.
5 Fit the new windscreen with a new rubber surround. Reusing the old one, even though it may look alright, is likely to result in leaks. Roll the rubber onto the glass, and get it seating closely all round.
6 At this stage the glass should be lying on something like a blanket, outside downwards. It could easily get scratched when on the ground.
7 Into the channel on the outside of the rubber surround thread a cord. This wants to be thin, but very strong. Terylene from a yacht chandler is ideal. It wants to be about 16 ft (5 metres) long. The two ends of the cord should overlap each other by about a foot, brought together at the bottom.
8 Get two assistants to lift the glass into place, putting the cords through the aperture into the car. Now steadily draw the cords to pull the lip of the rubber surround over the sill and into place. When the wire running up to the light in the mirror is reached this must be fed in too.
9 From outside work round the outer sealing lip of the rubber surround bending it back and injecting sealing compound underneath.
10 One last word. The professional will fit the screen, with sealant round the joint without making a mess, quickly, and therefore cheaply. This is a job for a garage.

9 Back window and fixed side window glasses

1 The back window glasses, and the fixed ones in the side, are dealt with in the same way as the windscreen.
2 These glasses are likely to be toughened glass, not laminated, so easier to fit without risk of cracking.

10 Door windows

1 The sliding windoes in the driver's and passenger's doors are controlled by an arm connected by a train of gears to the handle.
2 To gain access for the removal of the mechanism or the glass, remove the door trim panel as described later in section 12.

3 An assistant will be a help to control the glass when the winding mechanism has been removed, (photo).
4 Undo the bolt securing the front guide channel for the glass to the winding mechanism.
5 Remove the three bolts holding the mechanism to the door stiffening lining. The one shown in out photos has been superceded by more comprehensive door interior, and mountings for the mechanism less widely spread.
6 Slide the mechanism clear of the door panels, and also forward, so the boss on its actuating arm can be slid out of the groove in the metal channel on the bottom edge of the glass, (photo).
7 With the winder mechanism out of the way, the front glass channel can be taken out, and then the glass worked out of the door panel, (photo).
8 Before refitting make sure the lining of all the channels are in good condition, and also that there is no rust in the cavity behind the channels. Lubricate the winder, including its slider at the bottom of the glass.
9 When refitting, make sure the mechanism works freely and adjust the mounting for the front channel to hold the glass snugly, but not so tight it is stiff.

11 Door ventilators

1 If a new glass has to be fitted, get a new rubber sealing strip as well. Fit the rubber to the glass. Then wet its outside with petrol as a lubricant, and slide it into the frame.
2 To remove the frame from the door, the mounting pin at the top of the ventilator must be removed, br drilling it out. The drill should come from beneath. A new pin will be needed, or a chrome or brass screw with washers and self lock nut, (photo).

12 Door lining trim panel

1 The panels must be removed to gain access to the window mechanism, or to check for rust inside the doors.
2 Push the panel in, away from the window winder, to reveal the securing device for the handle. This is a little kidney shaped key, held in place by the boss of the plastic escutcheon behind the handle. Remove the key, and take off the handle, (photo).
3 On later cars again press in the panel, to free the door handle.
4 On early cars the trim panel is secured to the door by tabs bent over the edges, and a groove at the bottom. Bend the tabs clear, and then pull the panel at the sides, so that it bends, and bursts out of the groove, (photo).
5 Later cars have spring clips. Try pulling the panel out a bit to see just where a clip is. Then insert a screwdriver near the clip, and carefully prise it out. Then work along doing the others.
6 Whenever the panel is off, check for rust in the doors, and that the drain holes are clear. A thick coat of underneath paint will keep away rust, and sound proof the car.

13 Door rattles - tracing and rectification

Door rattles are due either to loose hinges, worn or maladjusted catches, or loose components inside the door. Loose hinges can be detected by opening the door and trying to lift it. Any play will be felt. Worn or badly adjusted catches can be found by pushing and pulling on the outside handle when the door is closed. Once again any play will be felt. To check the window mechanism open the door and shake it with the window first open and then closed.

14 Door hinges

1 As the hinges wear, the outer end of the door by the lock will get lower. The striker plate has to hold it up to keep it in the correct position. Not only would it look unsightly if drooping

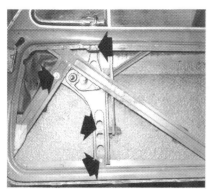

10.3. The channel must be released from the winder, and the winder mechanism from the door

10.6. Take out the winder, supporting the glass

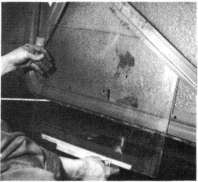

10.7. Then the glass can be lowered by hand, and taken out

11.2. To remove the ventilator, drill out the top pivot from below

12.2. The window winder kidney shaped retaining key is released by pressing the trim panel and plastic escutcheon towards the door

12.4. Some cars have spring clips. This one has tabs that must be unbent. Then the panel is curved and pulled out of the lips top and bottom

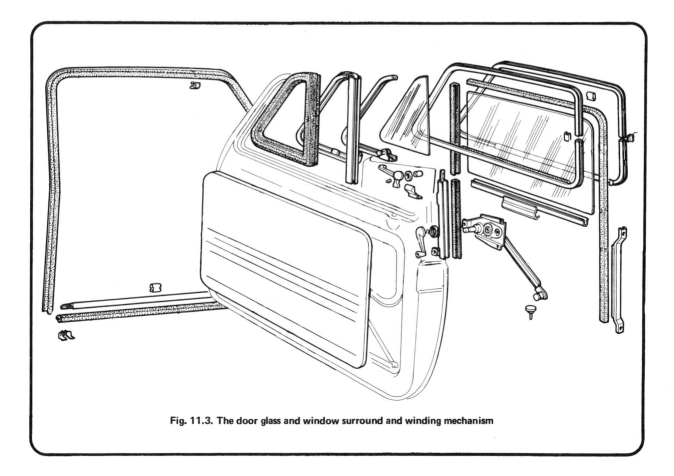

Fig. 11.3. The door glass and window surround and winding mechanism

and also allow in wet or wind around the seal, but it would also foul the sill.

2 As the striker has to lift the door, its closing becomes stiff as the car grows old. In the end it may be necessary to renovate the hinges.

3 The door half of the hinge is welded on. So a full replacement is impractical. New pins might be available, to special order, but a local agent is unlikely to stock them.

4 Remove the door, by unbolting the recessed mounting for the pin from the body. Then dismantle the hinge. The pin should be driven out upwards. It will probably need a good soak in rust remover. In difficult cases it may need drilling the whole way.

5 Once the old pin is out the hinges can be renovated successfully. Get some very long bolts, so that the shanks between the end of the thread and the head can be used as hinge pins. Ideally the hole through the hinge should be reamed, but a second best is to drill taking great care to get it straight. If the hinge is badly worn on the edges as well, put washers between the two halves to take the weight, and restore the doors to their original height. Grease it all thoroughly before assembly.

6 The fixture of the bolts depends on the type of hinge fitted, as these vary. On the early ones there will be room to fit a nut and lock nut, keeping the bolt as such. On others there will not. Carefully check just how much room your particular car has where the hinge is recessed into the body. There may be space for the head of the bolt on top, but not for nuts below. If there is no other way the end of the pin must be burred over. Cut it off leaving about 1/8 in (3 mm) spare. Hammer the end till it spreads to form a head.

15 Door and luggage locker lid fit

1 The fit of all doors is important. On original assembly what is quite a tricky job was done in a moment by someone used to it.
2 When taking off any door hinge, or the door lock striker plates, mark their position with pencil so that they can be put back in the same place, or if adjustments are being made, the starting point is known.

16 Emergency access to luggage locker

1 Should the flexible cable to the catch for the locker lid break, it is difficult to free the catch, as the locker is without other entrance.
2 If this happens, remove a headlight unit. If you are right handed, the car's left one is probably the best. On cars with the early type secured by the press and turn action of the adjusters in slots, this is not too difficult. On later cars it may mean breaking the light to get it out.
3 With the light unit out of the way put an arm in through the light unit's hole, and reach round to the lever for the catch in the centre.
4 On no account try prising at the joint between lid and body. If the attack through the light fails, it is better to cut the lid panel near the catch.
5 Drill a hole to start it. From there make a "V" cut and peel back the metal to make a triangular hole. This is easier to mend than a deformed joint.

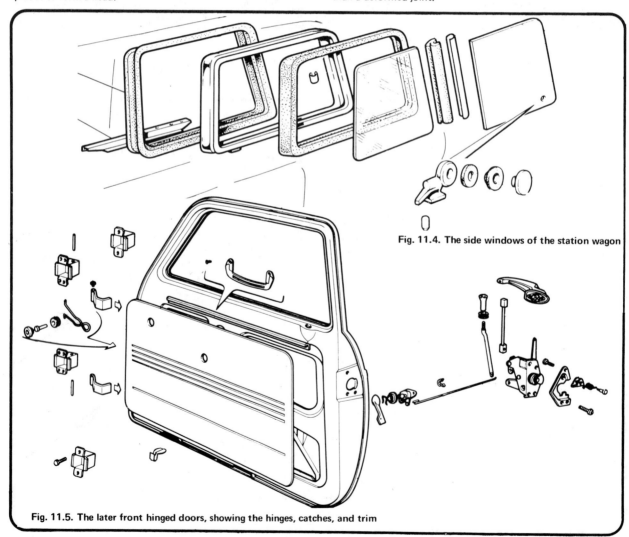

Fig. 11.4. The side windows of the station wagon

Fig. 11.5. The later front hinged doors, showing the hinges, catches, and trim

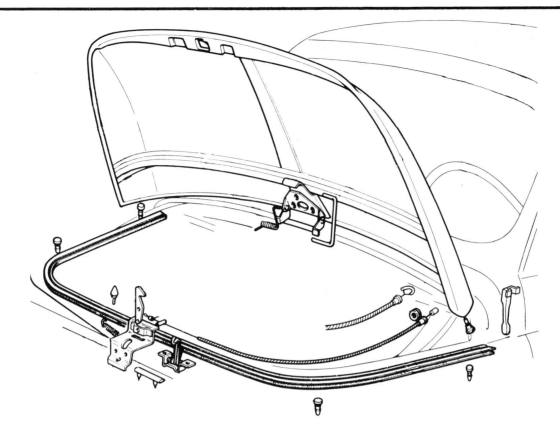

Fig. 11.6. The luggage locker, or boot, lid and catch arrangement

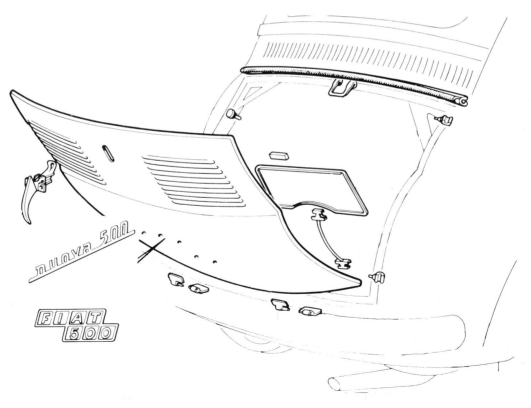

Fig. 11.7. Engine cover, hinges and catch (sedan)

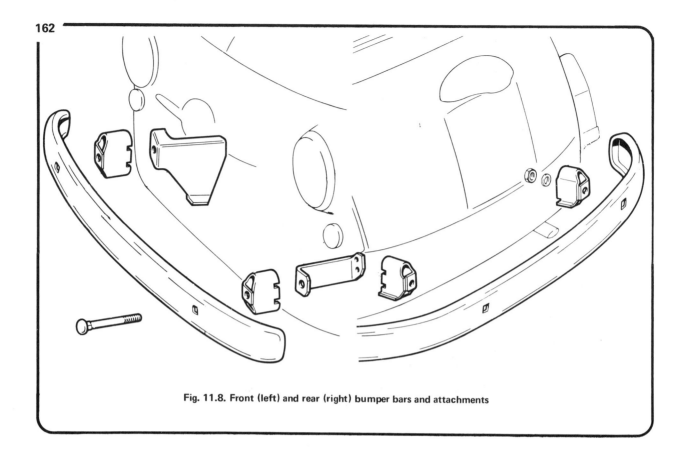

Fig. 11.8. Front (left) and rear (right) bumper bars and attachments

Metric conversion tables

Inches	Decimals	Millimetres	Millimetres to Inches		Inches to Millimetres	
			mm	Inches	Inches	mm
1/64	0.015625	0.3969	0.01	0.00039	0.001	0.0254
1/32	0.03125	0.7937	0.02	0.00079	0.002	0.0508
3/64	0.046875	1.1906	0.03	0.00118	0.003	0.0762
1/16	0.0625	1.5875	0.04	0.00157	0.004	0.1016
5/64	0.078125	1.9844	0.05	0.00197	0.005	0.1270
3/32	0.09375	2.3812	0.06	0.00236	0.006	0.1524
7/64	0.109375	2.7781	0.07	0.00276	0.007	0.1778
1/8	0.125	3.1750	0.08	0.00315	0.008	0.2032
9/64	0.140625	3.5719	0.09	0.00354	0.009	0.2286
5/32	0.15625	3.9687	0.1	0.00394	0.01	0.254
11/64	0.171875	4.3656	0.2	0.00787	0.02	0.508
3/16	0.1875	4.7625	0.3	0.01181	0.03	0.762
13/64	0.203125	5.1594	0.4	0.01575	0.04	1.016
7/32	0.21875	5.5562	0.5	0.01969	0.05	1.270
15/64	0.234375	5.9531	0.6	0.02362	0.06	1.524
1/4	0.25	6.3500	0.7	0.02756	0.07	1.778
17/64	0.265625	6.7469	0.8	0.03150	0.08	2.032
9/32	0.28125	7.1437	0.9	0.03543	0.09	2.286
19/64	0.296875	7.5406	1	0.03937	0.1	2.54
5/16	0.3125	7.9375	2	0.07874	0.2	5.08
21/64	0.328125	8.3344	3	0.11811	0.3	7.62
11/32	0.34375	8.7312	4	0.15748	0.4	10.16
23/64	0.359375	9.1281	5	0.19685	0.5	12.70
3/8	0.375	9.5250	6	0.23622	0.6	15.24
25/64	0.390625	9.9219	7	0.27559	0.7	17.78
13/32	0.40625	10.3187	8	0.31496	0.8	20.32
27/64	0.421875	10.7156	9	0.35433	0.9	22.86
7/16	0.4375	11.1125	10	0.39370	1	25.4
29/64	0.453125	11.5094	11	0.43307	2	50.8
15/32	0.46875	11.9062	12	0.47244	3	76.2
31/64	0.484375	12.3031	13	0.51181	4	101.6
1/2	0.5	12.7000	14	0.55118	5	127.0
33/64	0.515625	13.0969	15	0.59055	6	152.4
17/32	0.53125	13.4937	16	0.62992	7	177.8
35/64	0.546875	13.8906	17	0.66929	8	203.2
9/16	0.5625	14.2875	18	0.70866	9	228.6
37/64	0.578125	14.6844	19	0.74803	10	254.0
19/32	0.59375	15.0812	20	0.78740	11	279.4
39/64	0.609375	15.4781	21	0.82677	12	304.8
5/8	0.625	15.8750	22	0.86614	13	330.2
41/64	0.640625	16.2719	23	0.90551	14	355.6
21/32	0.65625	16.6687	24	0.94488	15	381.0
43/64	0.671875	17.0656	25	0.98425	16	406.4
11/16	0.6875	17.4625	26	1.02362	17	431.8
45/64	0.703125	17.8594	27	1.06299	18	457.2
23/32	0.71875	18.2562	28	1.10236	19	482.6
47/64	0.734375	18.6531	29	1.14173	20	508.0
3/4	0.75	19.0500	30	1.18110	21	533.4
49/64	0.765625	19.4469	31	1.22047	22	558.8
25/32	0.78125	19.8437	32	1.25984	23	584.2
51/64	0.796875	20.2406	33	1.29921	24	609.6
13/16	0.8125	20.6375	34	1.33858	25	635.0
53/64	0.828125	21.0344	35	1.37795	26	660.4
27/32	0.84375	21.4312	36	1.41732	27	685.8
55/64	0.859375	21.8281	37	1.4567	28	711.2
7/8	0.875	22.2250	38	1.4961	29	736.6
57/64	0.890625	22.6219	39	1.5354	30	762.0
29/32	0.90625	23.0187	40	1.5748	31	787.4
59/64	0.921875	23.4156	41	1.6142	32	812.8
15/16	0.9375	23.8125	42	1.6535	33	838.2
61/64	0.953125	24.2094	43	1.6929	34	863.6
31/32	0.96875	24.6062	44	1.7323	35	889.0
63/64	0.984375	25.0031	45	1.7717	36	914.4

Index

A
Air cleaner - 68
Air control flaps - heater - 65

B
Ball joints - steering - 144
Battery - maintenance - 119
Belt - fan - tension - 64
Big ends (see crankshaft)
Bodywork
    corrosion - 156
    doors - 158
    major and minor repairs - 155
    upholstery maintenance - 155
Braking system
    adjustment - 108
    back plates - 114
    bleeding - 108
    fault finding chart - 116
    front drums and shoes - 110
    general description - 107
    handbrake - 113
    hydraulic lines - 108
    master cylinder - 112
    rear drums - 110
    rear shoes - 110
    specifications - 107
    wheel cylinders - 111
Brushes - generator - 122
Brushes - starter motor - 126
Bulbs - lamps - 130
Bumpers - 162

C
Cam followers - (see camshaft)
Camshaft - 35, 47
Carburettor
    adjustments - 71
    description - 67
    removal, dismantling and replacement - 68
Clutch
    cable operating adjustment - 86
    fault finding chart - 88
    release operating mechanism - 84
    removal, inspection and reassembly - 84
Coil - testing 80
Condenser - testing 80
Connecting rod - 35, 47
Contact breaker
    adjustment - 77
Cooling system
    fan belt adjustment - 64
    fan removal and replacement - 62
    general description - 59
    heater controls - 65
    thermostat - 65
Crankshaft - 38, 40, 45
Cut-out - 124
Cylinder head - 28, 34, 43, 47
Cylinder head decarbonisation - 43
Cylinders 40, 45

D
Decarbonisation - 43
Deflector plates - engine - 64
Direction indicators - 130
Distributor
    contact breaker points - 77
    dismantling, overhaul and reassembly - 78
    static ignition timing - 77
Door hinges - 158
Doors - 158
Door window - 158
Drive shafts -
Drum brakes
    front - 110
    rear - 110
Dynamo - 120

E
Electrical system
    battery - 119
    battery charging - 119
    control box - 124
    direction indicators - 130
    fault finding - 132
    fuses - 128
    general description - 119
    generator - removal and replacement - 122
    generator - testing - 120
    horn - 130
    lamps - 130
    specifications - 117
    starter motor - testing - 126
    starter motor - overhaul - 126
    stop lamps - 130
    windscreen wipers - 129
    wiring diagrams - 134
Engine
    camshaft - 35, 47
    centrifugal oil filter - 35, 47
    connecting rods - 35, 47
    crankcase - 35, 47
    crankshaft - 38, 40, 45
    cylinder head - 28, 34, 43, 47
    cylinders - 40, 45
    decarbonising - 43
    dismantling - 31

Index

fault finding - 52
flywheel - 35, 45, 47
gudgeon pins - 35, 40
oil pump - 38, 45, 47
piston rings - 40, 47
pistons - 35, 40, 47
reassembly - final - 50, 52
removal - 31
replacement - 50
rocker gear - 44
specifications - 19
static timing - 77
timing chain - 35, 47
valve gear - 44
Exhaust system - 66

F

Fan - 62
Fan belt adjustment - 64
Flywheel - 24, 30, 42
Front hub bearings - 144
Front suspension - 143
Front suspension setting angles - 148
Fuel pump - 72
Fuel system
    air cleaner - 67
    carburettor - 67
    fault finding - 74
    fuel pump - 72
    general description - 67
    specifications - 67

G

Gearbox (see Transmission)
Generator
    removal and replacement - 122
    testing - 120
Gudgeon pins - 35, 40

H

Handbrake - 113
Headlamps - 130
Headlamps - beam setting - 130
Heater controls - 65
Horn - 130
Hub bearing (front) - 144
Hub bearings (rear) - 138
Hydraulic brake operation - 107
Hydraulic lines - 108
Hydrometer test - electrolyte - 119

I

Ignition
    coil - 80
    distributor - 78
    fault finding - 80
    general description - 75
    spark plugs - 80
    specifications - 75
    static engine timing - 77
Instruments - 150

L

Locks - 158
Lubrication
    chart - 15
    engine system - 59
    recommended grades - 16

M

Master cylinder - 112

O

Oil cooling duct - 59
Oil pump - 38, 45, 47

P

Piston rings - 35, 40, 47
Pistons - 35, 40, 47
Points - contact breaker - 77
Pump - fuel - 72
Pushrods and cam followers (see camshaft)

R

Rear brakes - 110
Rear lamps - 130
Regulator - 124
Removal and replacement of engine - 31, 50
Rocker adjustment and clearances - 8
Rocker assembly - 144
Routine maintenance - 7
Rear suspension - 137

S

Shock absorbers - 142
Silencer - 66
Spark plugs - 80
Specific gravity - electrolyte - 119
Starter motor
    overhaul - 126
    testing - 126
Steering
    adjustments - 150
    description - 150
    dismantling and overhaul - 152
    fault finding - 154
    king pins - 147
    removal and replacement - 150
    specifications - 143
    track rod and joints - 144
    wheel and column - 152
Suspension
    dampers - 142
    dismantling front - 144
    dismantling rear - 138
    geometry rear - 140
    specifications - front - 143
    specifications - rear - 137
Suspension pivots - 148
    front wheel bearings - 144
    Wheels and tyres - 142

T

Thermostat - 65
Timing marks - 77
Transmission
    differential bearings - 93
    dismantling - 93
    fault diagnosis - 106
    differential meshing - 100
    drive shafts - 93
    general description - 90
    input shaft assembly - 103
    input shaft oil seal - 103
    main casing bearing assembly - 103
    output shaft assembly - 103
    reassembly - gearbox - 103
    removal - gearbox - 90
    replacement - 105
    reverse gear shaft - 103
    specification - 89

Index

Trim - door - removal - 158
Tyres - 142

V

Valve tappet clearances - 8
Voltage regulator - 124

W

Wheels and tyres - 142
Window glass - 158
Windscreen glass - 158
Windscreen wipers - 129
Wiring diagrams - 134

Printed by
J. H. HAYNES & Co. Ltd
Sparkford Yeovil Somerset